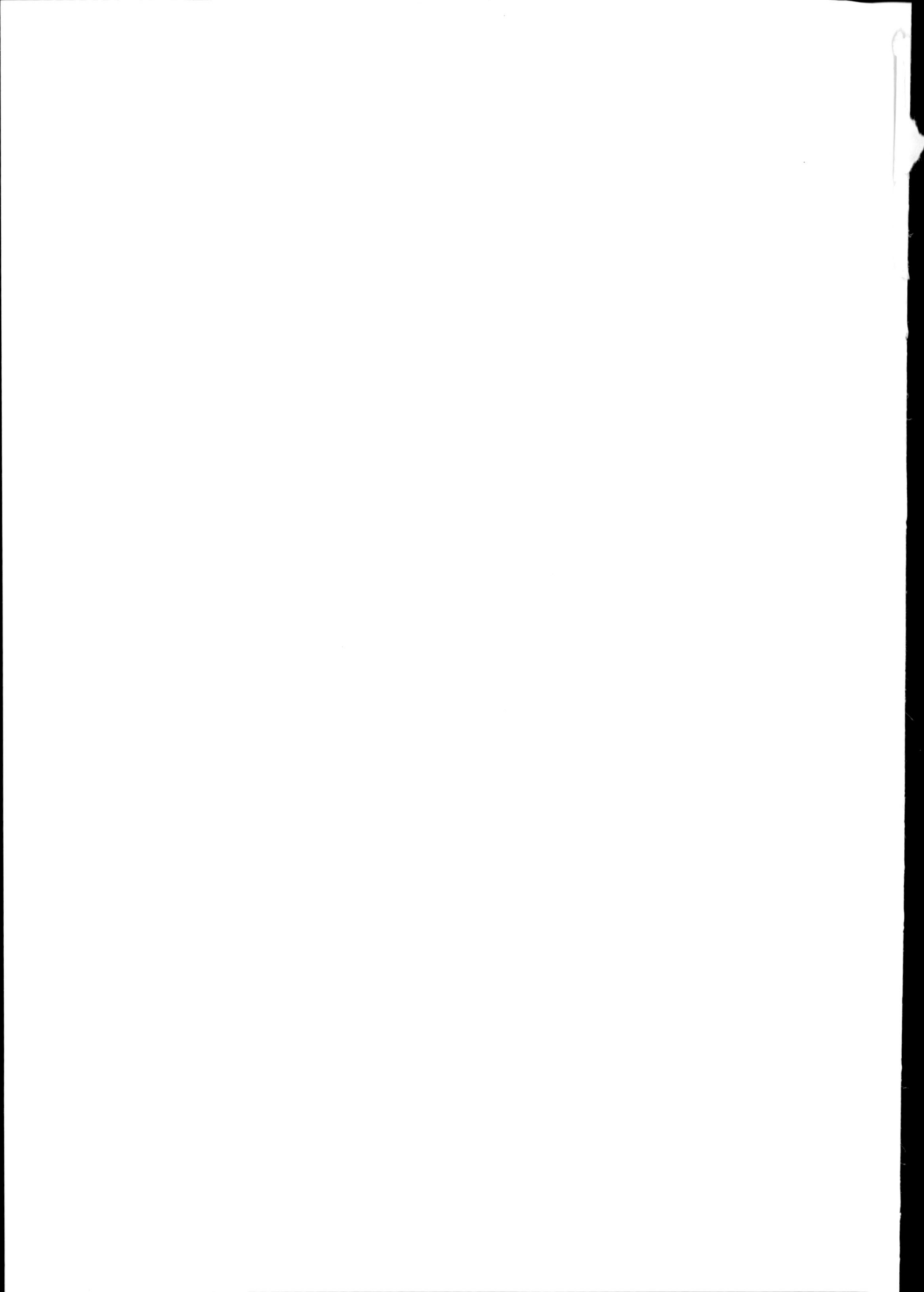

An Introduction to Astrophysics

Edited by
Mervin Williamson

Larsen & Keller
www.larsen-keller.com

An Introduction to Astrophysics
Edited by Mervin Williamson
ISBN: 978-1-63549-034-3 (Hardback)

☰ Larsen & Keller

Published by Larsen and Keller Education,
5 Penn Plaza,
19th Floor,
New York, NY 10001, USA

Cataloging-in-Publication Data

An introduction to astrophysics / edited by Mervin Williamson.
    p. cm.
Includes bibliographical references and index.
ISBN 978-1-63549-034-3
1. Astrophysics. 2. Astronomy. 3. Stars. 4. Black holes (Astronomy). I. Williamson, Mervin.
QB461 .I58 2017
523.01--dc23

The publisher's policy is to use permanent paper from mills that operate a sustainable forestry policy. Furthermore, the publisher ensures that the text paper and cover boards used have met acceptable environmental accreditation standards.

Printed and bound in the United States of America.

For more information regarding Larsen and Keller Education and its products, please visit the publisher's website www.larsen-keller.com

Table of Contents

Preface

As a part of astronomy, astrophysics deals with the study of celestial bodies by applying the laws of physics and chemistry. The field applies laws of different subjects like thermodynamics, molecular physics, relativity, particle physics, etc. This book unfolds the innovative aspects of astrophysics, which will be crucial for the holistic understanding of the subject matter. The topics included in it are of utmost significance are and bound to provide incredible insights to the readers. This textbook is meant for students who are looking for an elaborate reference text on astrophysics. The book aims to shed light on some of the unexplored aspects of astrophysics.

A short introduction to every chapter is written below to provide an overview of the content of the book:

Chapter 1 - This chapter will provide an integrated understanding of astrophysics. The branch of astronomy that employs the principles of physics and chemistry to study space and interstellar bodies is astrophysics. The content is an overview of the subject matter incorporating all the major aspects of astrophysics; **Chapter 2 -** The major approaches of astrophysics are discussed in this chapter. Radio astronomy and high-energy astronomy are significant and important topics related to astrophysics. High-energy astronomy is further elaborated with the help of X-ray astronomy, gamma-ray astronomy and ultraviolet astronomy. Major components and keys concepts of astrophysics are strategically incorporated in this chapter; **Chapter 3 -** Astrophysics is best understood in confluence with the major topics listed in the following chapter. The major concepts of astrophysics such as nebula, active galactic nucleus, Hayashi track, and the Eddington number are explained. This chapter serves as a source to understand the major concepts related to astrophysics; **Chapter 4 -** A star is largely composed of gases and it produces heat, light, ultraviolent rays and other forms of radiations. The closest star to the earth is the Sun. This chapter discusses star formation, stellar evolution, supernova, and white dwarf. It elucidates the crucial theories and principles of stars; **Chapter 5 -** The major components of black holes are discussed in this chapter. The chapter strategically encompasses and incorporates the major components and key concepts of black holes, such as, black hole starship, black hole information paradox and black hole complementarity.

I extend my sincere thanks to the publisher for considering me worthy of this task. Finally, I thank my family for being a source of support and help.

Editor

Introduction to Astrophysics

This chapter will provide an integrated understanding of astrophysics. The branch of astronomy that employs the principles of physics and chemistry to study space and interstellar bodies is astrophysics. The content is an overview of the subject matter incorporating all the major aspects of astrophysics.

Astrophysics is the branch of astronomy that employs the principles of physics and chemistry "to ascertain the nature of the heavenly bodies, rather than their positions or motions in space." Among the objects studied are the Sun, other stars, galaxies, extrasolar planets, the interstellar medium and the cosmic microwave background. Their emissions are examined across all parts of the electromagnetic spectrum, and the properties examined include luminosity, density, temperature, and chemical composition. Because astrophysics is a very broad subject, *astrophysicists* typically apply many disciplines of physics, including mechanics, electromagnetism, statistical mechanics, thermodynamics, quantum mechanics, relativity, nuclear and particle physics, and atomic and molecular physics.

In practice, modern astronomical research often involves a substantial amount of work in the realms of theoretical and observational physics. Some areas of study for astrophysicists include their attempts to determine: the properties of dark matter, dark energy, and black holes; whether or not time travel is possible, wormholes can form, or the multiverse exists; and the origin and ultimate fate of the universe. Topics also studied by theoretical astrophysicists include: Solar System formation and evolution; stellar dynamics and evolution; galaxy formation and evolution; magnetohydrodynamics; large-scale structure of matter in the universe; origin of cosmic rays; general relativity and physical cosmology, including string cosmology and astroparticle physics.

Astrophysics can be studied at the bachelors, masters, and Ph.D. levels in physics or astronomy departments at many universities.

History

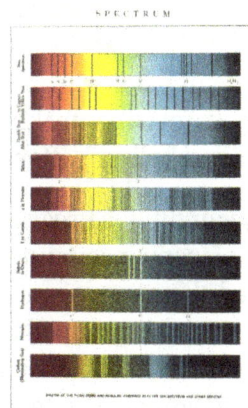

Early 20th-century comparison of elemental, solar, and stellar spectra

Although astronomy is as ancient as recorded history itself, it was long separated from the study of terrestrial physics. In the Aristotelian worldview, bodies in the sky appeared to be unchanging spheres whose only motion was uniform motion in a circle, while the earthly world was the realm which underwent growth and decay and in which natural motion was in a straight line and ended when the moving object reached its goal. Consequently, it was held that the celestial region was made of a fundamentally different kind of matter from that found in the terrestrial sphere; either Fire as maintained by Plato, or Aether as maintained by Aristotle. During the 17th century, natural philosophers such as Galileo, Descartes, and Newton began to maintain that the celestial and terrestrial regions were made of similar kinds of material and were subject to the same natural laws. Their challenge was that the tools had not yet been invented with which to prove these assertions.

Although astronomy is as ancient as recorded history itself, it was long separated from the study of terrestrial physics. In the Aristotelian worldview, bodies in the sky appeared to be unchanging spheres whose only motion was uniform motion in a circle, while the earthly world was the realm which underwent growth and decay and in which natural motion was in a straight line and ended when the moving object reached its goal. Consequently, it was held that the celestial region was made of a fundamentally different kind of matter from that found in the terrestrial sphere; either Fire as maintained by Plato, or Aether as maintained by Aristotle. During the 17th century, natural philosophers such as Galileo, Descartes, and Newton began to maintain that the celestial and terrestrial regions were made of similar kinds of material and were subject to the same natural laws. Their challenge was that the tools had not yet been invented with which to prove these assertions.

For much of the nineteenth century, astronomical research was focused on the routine work of measuring the positions and computing the motions of astronomical objects. A new astronomy, soon to be called astrophysics, began to emerge when William Hyde Wollaston and Joseph von Fraunhofer independently discovered that, when decomposing the light from the Sun, a multitude of dark lines (regions where there was less or no light) were observed in the spectrum. By 1860 the physicist, Gustav Kirchhoff, and the chemist, Robert Bunsen, had demonstrated that the dark lines in the solar spectrum corresponded to bright lines in the spectra of known gases, specific lines corresponding to unique chemical elements. Kirchhoff deduced that the dark lines in the solar spectrum are caused by absorption by chemical elements in the Solar atmosphere. In this way it was proved that the chemical elements found in the Sun and stars were also found on Earth.

Among those who extended the study of solar and stellar spectra was Norman Lockyer, who in 1868 detected bright, as well as dark, lines in solar spectra. Working with the chemist, Edward Frankland, to investigate the spectra of elements at various temperatures and pressures, he could not associate a yellow line in the solar spectrum with any known elements. He thus claimed the line represented a new element, which was called helium, after the Greek Helios, the Sun personified.

In 1885, Edward C. Pickering undertook an ambitious program of stellar spectral classification at Harvard College Observatory, in which a team of woman computers, notably Williamina Fleming, Antonia Maury, and Annie Jump Cannon, classified the spectra recorded on photographic plates. By 1890, a catalog of over 10,000 stars had been prepared that grouped them into thirteen spectral types. Following Pickering's vision, by 1924 Cannon expanded the catalog to nine volumes and over a quarter of a million stars, developing the Harvard Classification Scheme which was accepted for world-wide use in 1922.

In 1895, George Ellery Hale and James E. Keeler, along with a group of ten associate editors from Europe and the United States, established The Astrophysical Journal: An International Review of Spectroscopy and Astronomical Physics. It was intended that the journal would fill the gap between journals in astronomy and physics, providing a venue for publication of articles on astronomical applications of the spectroscope; on laboratory research closely allied to astronomical physics, including wavelength determinations of metallic and gaseous spectra and experiments on radiation and absorption; on theories of the Sun, Moon, planets, comets, meteors, and nebulae; and on instrumentation for telescopes and laboratories.

In 1925 Cecilia Helena Payne (later Cecilia Payne-Gaposchkin) wrote an influential doctoral dissertation at Radcliffe College, in which she applied ionization theory to stellar atmospheres to relate the spectral classes to the temperature of stars. Most significantly, she discovered that hydrogen and helium were the principal components of stars. This discovery was so unexpected that her dissertation readers convinced her to modify the conclusion before publication. However, later research confirmed her discovery.

By the end of the 20th century, further study of stellar and experimental spectra advanced, particularly as a result of the advent of quantum physics.

Observational Astrophysics

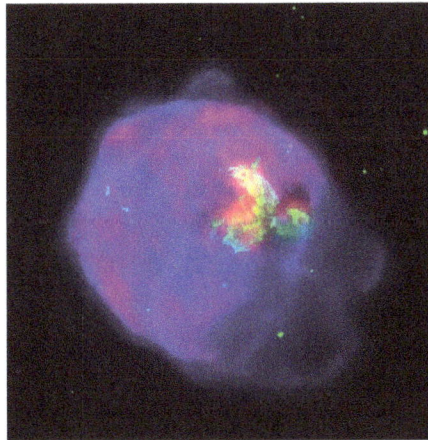

Supernova remnant LMC N 63A imaged in x-ray (blue), optical (green) and radio (red) wavelengths. The X-ray glow is from material heated to about ten million degrees Celsius by a shock wave generated by the supernova explosion.

Observational astronomy is a division of the astronomical science that is concerned with recording data, in contrast with theoretical astrophysics, which is mainly concerned with finding out the measurable implications of physical models. It is the practice of observing celestial objects by using telescopes and other astronomical apparatus.

The majority of astrophysical observations are made using the electromagnetic spectrum.

- Radio astronomy studies radiation with a wavelength greater than a few millimeters. Example areas of study are radio waves, usually emitted by cold objects such as interstellar gas and dust clouds; the cosmic microwave background radiation which is the redshifted light from the Big Bang; pulsars, which were first detected at microwave frequencies. The study of these waves requires very large radio telescopes.

- Infrared astronomy studies radiation with a wavelength that is too long to be visible to the naked eye but is shorter than radio waves. Infrared observations are usually made with telescopes similar to the familiar optical telescopes. Objects colder than stars (such as planets) are normally studied at infrared frequencies.

- Optical astronomy is the oldest kind of astronomy. Telescopes paired with a charge-coupled device or spectroscopes are the most common instruments used. The Earth's atmosphere interferes somewhat with optical observations, so adaptive optics and space telescopes are used to obtain the highest possible image quality. In this wavelength range, stars are highly visible, and many chemical spectra can be observed to study the chemical composition of stars, galaxies and nebulae.

- Ultraviolet, X-ray and gamma ray astronomy study very energetic processes such as binary pulsars, black holes, magnetars, and many others. These kinds of radiation do not penetrate the Earth's atmosphere well. There are two methods in use to observe this part of the electromagnetic spectrum—space-based telescopes and ground-based imaging air Cherenkov telescopes (IACT). Examples of Observatories of the first type are RXTE, the Chandra X-ray Observatory and the Compton Gamma Ray Observatory. Examples of IACTs are the High Energy Stereoscopic System (H.E.S.S.) and the MAGIC telescope.

Other than electromagnetic radiation, few things may be observed from the Earth that originate from great distances. A few gravitational wave observatories have been constructed, but gravitational waves are extremely difficult to detect. Neutrino observatories have also been built, primarily to study our Sun. Cosmic rays consisting of very high energy particles can be observed hitting the Earth's atmosphere.

Observations can also vary in their time scale. Most optical observations take minutes to hours, so phenomena that change faster than this cannot readily be observed. However, historical data on some objects is available, spanning centuries or millennia. On the other hand, radio observations may look at events on a millisecond timescale (millisecond pulsars) or combine years of data (pulsar deceleration studies). The information obtained from these different timescales is very different.

The study of our very own Sun has a special place in observational astrophysics. Due to the tremendous distance of all other stars, the Sun can be observed in a kind of detail unparalleled by any other star. Our understanding of our own Sun serves as a guide to our understanding of other stars.

The topic of how stars change, or stellar evolution, is often modeled by placing the varieties of star types in their respective positions on the Hertzsprung–Russell diagram, which can be viewed as representing the state of a stellar object, from birth to destruction.

Theoretical Astrophysics

Theoretical astrophysicists use a wide variety of tools which include analytical models (for example, polytropes to approximate the behaviors of a star) and computational numerical simulations. Each has some advantages. Analytical models of a process are generally better for giving insight into the heart of what is going on. Numerical models can reveal the existence of phenomena and effects that would otherwise not be seen.

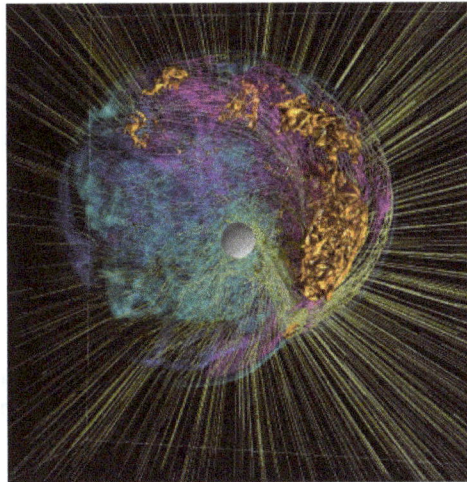

Stream lines on this simulation of a supernova show the flow of matter behind the shock wave giving clues as to the origin of pulsars

Theorists in astrophysics endeavor to create theoretical models and figure out the observational consequences of those models. This helps allow observers to look for data that can refute a model or help in choosing between several alternate or conflicting models.

Theorists also try to generate or modify models to take into account new data. In the case of an inconsistency, the general tendency is to try to make minimal modifications to the model to fit the data. In some cases, a large amount of inconsistent data over time may lead to total abandonment of a model.

Topics studied by theoretical astrophysicists include: stellar dynamics and evolution; galaxy formation and evolution; magnetohydrodynamics; large-scale structure of matter in the universe; origin of cosmic rays; general relativity and physical cosmology, including string cosmology and astroparticle physics. Astrophysical relativity serves as a tool to gauge the properties of large scale structures for which gravitation plays a significant role in physical phenomena investigated and as the basis for black hole (*astro*)physics and the study of gravitational waves.

Some widely accepted and studied theories and models in astrophysics, now included in the Lambda-CDM model, are the Big Bang, cosmic inflation, dark matter, dark energy and fundamental theories of physics. Wormholes are examples of hypotheses which are yet to be proven (or disproven).

Popularization

The roots of astrophysics can be found in the seventeenth century emergence of a unified physics, in which the same laws applied to the celestial and terrestrial realms. There were scientists who were qualified in both physics and astronomy who laid the firm foundation for the current science of astrophysics. In modern times, students continue to be drawn to astrophysics due to its popularization by the Royal Astronomical Society and notable educators such as prominent professors Subrahmanyan Chandrasekhar, Stephen Hawking, Hubert Reeves, Carl Sagan and Neil deGrasse Tyson. The efforts of the early, late, and present scientists continue to attract young people to study the history and science of astrophysics.

References

- Lloyd, G.E.R. (1968). Aristotle: The Growth and Structure of His Thought. Cambridge: Cambridge University Press. pp. 134–5. ISBN 0-521-09456-9.

- Galilei, Galileo (1989), Van Helden, Albert, ed., Sidereus Nuncius or The Sidereal Messenger, Chicago: University of Chicago Press, pp. 21, 47, ISBN 0-226-27903-0

- Westfall, Richard S. (1980), Never at Rest: A Biography of Isaac Newton, Cambridge: Cambridge University Press, pp. 731–732, ISBN 0-521-27435-4

- Burtt, Edwin Arthur (2003) [First published 1924], The Metaphysical Foundations of Modern Science (second revised ed.), Mineola, NY: Dover Publications, pp. 30, 41, 241–2, ISBN 9780486425511

- Haramundanis, Katherine (2007), "Payne-Gaposchkin [Payne], Cecilia Helena", in Hockey, Thomas; Trimble, Virginia; Williams, Thomas R., Biographical Encyclopedia of Astronomers, New York: Springer, pp. 876–878, ISBN 978-0-387-30400-7, retrieved July 19, 2015

- Ladislav Kvasz (2013). "Galileo, Descartes, and Newton – Founders of the Language of Physics" (PDF). Institute of Philosophy, Academy of Sciences of the Czech Republic. Retrieved 2015-07-18.

- Hetherington, Norriss S.; McCray, W. Patrick, Weart, Spencer R., ed., Spectroscopy and the Birth of Astrophysics, American Institute of Physics, Center for the History of Physics, retrieved July 19, 2015

2

Key Approaches to Astrophysics

The major approaches of astrophysics are discussed in this chapter. Radio astronomy and high-energy astronomy are significant and important topics related to astrophysics. High-energy astronomy is further elaborated with the help of X-ray astronomy, gamma-ray astronomy and ultraviolet astronomy. Major components and keys concepts of astrophysics are strategically incorporated in this chapter.

Radio Astronomy

The Very Large Array, a radio interferometer in New Mexico, USA

Radio astronomy is a subfield of astronomy that studies celestial objects at radio frequencies. The initial detection of radio waves from an astronomical object was made in the 1930s, when Karl Jansky observed radiation coming from the Milky Way. Subsequent observations have identified a number of different sources of radio emission. These include stars and galaxies, as well as entirely new classes of objects, such as radio galaxies, quasars, pulsars, and masers. The discovery of the cosmic microwave background radiation, regarded as evidence for the Big Bang theory, was made through radio astronomy.

Radio astronomy is conducted using large radio antennas referred to as radio telescopes, that are either used singularly, or with multiple linked telescopes utilizing the techniques of radio interferometry and aperture synthesis. The use of interferometry allows radio astronomy to achieve high angular resolution, as the resolving power of an interferometer is set by the distance between its components, rather than the size of its components.

History

Before Jansky observed the Milky Way in the 1930s, physicists speculated that radio waves could be observed from astronomical sources. In the 1860s, James Clerk Maxwell's equations had shown that electromagnetic radiation is associated with electricity and magnetism, and could exist at any wavelength. Several attempts were made to detect radio emission from the Sun including an experiment by German astrophysicists Johannes Wilsing and Julius Scheiner in 1896 and a centimeter wave radiation apparatus set up by Oliver Lodge between 1897-1900. These attempts were unable to detect any emission due to technical limitations of the instruments. The discovery of the radio reflecting ionosphere in 1902, led physicists to conclude that the layer would bounce any astronomical radio transmission back into space, making them undetectable.

Karl Jansky made the discovery of the first astronomical radio source serendipitously in the early 1930s. As an engineer with Bell Telephone Laboratories, he was investigating static that interfered with short wave transatlantic voice transmissions. Using a large directional antenna, Jansky noticed that his analog pen-and-paper recording system kept recording a repeating signal of unknown origin. Since the signal peaked about every 24 hours, Jansky originally suspected the source of the interference was the Sun crossing the view of his directional antenna. Continued analysis showed that the source was not following the 24-hour daily cycle of the Sun exactly, but instead repeating on a cycle of 23 hours and 56 minutes. Jansky discussed the puzzling phenomena with his friend, astrophysicist and teacher Albert Melvin Skellett, who pointed out that the time between the signal peaks was the exact length of a sidereal day, the timing found if the source was astronomical, "fixed" in relationship to the stars and passing in front of the antenna once every Earth rotation. By comparing his observations with optical astronomical maps, Jansky eventually concluded that the radiation source peaked when his antenna was aimed at the densest part of the Milky Way in the constellation of Sagittarius. He concluded that since the Sun (and therefore other stars) were not large emitters of radio noise, the strange radio interference may be generated by interstellar gas and dust in the galaxy. (Jansky's peak radio source, one of the brightest in the sky, was designated Sagittarius A in the 1950s and, instead of being galactic "gas and dust", has since be found to be emitted by electrons in a strong magnetic field from the complex of objects found in that area).

Jansky announced his discovery in 1933. He wanted to investigate the radio waves from the Milky Way in further detail, but Bell Labs reassigned him to another project, so he did no further work in the field of astronomy. His pioneering efforts in the field of radio astronomy have been recognized by the naming of the fundamental unit of flux density, the jansky (Jy), after him.

Grote Reber was inspired by Jansky's work, and built a parabolic radio telescope 9m in diameter in his backyard in 1937. He began by repeating Jansky's observations, and then conducted the first sky survey in the radio frequencies. On February 27, 1942, James Stanley Hey, a British Army research officer, made the first detection of radio waves emitted by the Sun. Later that year George Clark Southworth, at Bell Labs like Jansky, also detected radiowaves from the sun. Both researchers were bound by wartime security surrounding radar, so Reber, who was not, published his 1944 findings first. Several other people independently discovered solar radiowaves, including E. Schott in Denmark and Elizabeth Alexander working on Norfolk Island.

At Cambridge University, where ionospheric research had taken place during World War II, J.A. Ratcliffe along with other members of the Telecommunications Research Establishment that had carried out wartime research into radar, created a radiophysics group at the university where radio wave emissions from the Sun were observed and studied.

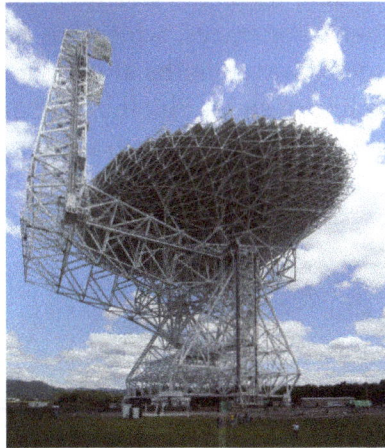

The Robert C. Byrd Green Bank Telescope (GBT) in West Virginia, United States is the world's largest fully steerable radio telescope.

This early research soon branched out into the observation of other celestial radio sources and interferometry techniques were pioneered to isolate the angular source of the detected emissions. Martin Ryle and Antony Hewish at the Cavendish Astrophysics Group developed the technique of Earth-rotation aperture synthesis. The radio astronomy group in Cambridge went on to found the Mullard Radio Astronomy Observatory near Cambridge in the 1950s. During the late 1960s and early 1970s, as computers (such as the Titan) became capable of handling the computationally intensive Fourier transform inversions required, they used aperture synthesis to create a 'One-Mile' and later a '5 km' effective aperture using the One-Mile and Ryle telescopes, respectively. They used the Cambridge Interferometer to map the radio sky, producing the famous 2C and 3C surveys of radio sources.

Techniques

First 7-metre ESO/NAOJ/NRAO ALMA Antenna.

Radio astronomers use different techniques to observe objects in the radio spectrum. Instruments may simply be pointed at an energetic radio source to analyze its emission. To "image" a region of the sky in more detail, multiple overlapping scans can be recorded and pieced together in a mosaic image. The type of instrument used depends on the strength of the signal and the amount of detail needed.

Observations from the Earth's surface are limited to wavelengths that can pass through the atmosphere. At low frequencies, or long wavelengths, transmission is limited by the ionosphere, which reflects waves with frequencies less than its characteristic plasma frequency. Water vapor interferes with radio astronomy at higher frequencies, which has led to building radio observatories that conduct observations at millimeter wavelengths at very high and dry sites, in order to minimize the water vapor content in the line of sight. Finally, transmitting devices on earth may cause radio-frequency interference. Because of this, many radio observatories are built at remote places.

Radio Telescopes

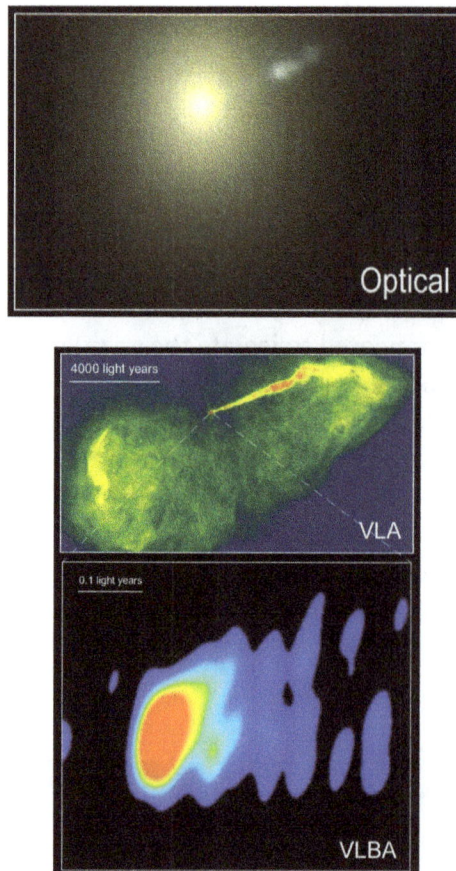

An optical image of the galaxy M87 (HST), a radio image of same galaxy using Interferometry (Very Large Array-VLA), and an image of the center section (VLBA) using a *Very Long Baseline Array* (Global VLBI) consisting of antennas in the US, Germany, Italy, Finland, Sweden and Spain. The jet of particles is suspected to be powered by a black hole in the center of the galaxy.

Radio telescopes may need to be extremely large in order to receive signals with high signal-to-noise ratio. Also since angular resolution is a function of the diameter of the "objective" in proportion to the wavelength of the electromagnetic radiation being observed, *radio telescopes* have to be much larger in comparison to their optical counterparts. For example, a 1-meter diameter optical telescope is two million times bigger than the wavelength of light observed giving it a resolution of roughly 0.3 arc seconds, whereas a radio telescope "dish" many times that size may, depending on the wavelength observed, only be able to resolve an object the size of the full moon (30 minutes of arc).

Radio Interferometry

The difficulty in achieving high resolutions with single radio telescopes led to radio interferometry, developed by British radio astronomer Martin Ryle and Australian engineer, radiophysicist, and radio astronomer Joseph Lade Pawsey and Ruby Payne-Scott in 1946. Surprisingly the first use of a radio interferometer for an astronomical observation was carried out by Payne-Scott, Pawsey and Lindsay McCready on 26 January 1946 using a SINGLE converted radar antenna (broadside array) at 200 MHz near Sydney, Australia. This group used the principle of a sea-cliff interferometer in which the antenna (formerly a World War II radar) observed the sun at sunrise with interference arising from the direct radiation from the sun and the reflected radiation from the sea. With this baseline of almost 200 meters, the authors determined that the solar radiation during the burst phase was much smaller than the solar disk and arose from a region associated with a large sunspot group. The Australia group laid out the principles of aperture synthesis in their ground-breaking paper submitted in mid-1946 and published in 1947. The use of a sea-cliff interferometer had been demonstrated by numerous groups in Australia, Iran and the UK during World War II, who had observed interference fringes (the direct radar return radiation and the reflected signal from the sea) from incoming aircraft.

The Cambridge group of Ryle and Vonberg observed the sun at 175 MHz for the first time in mid July 1946 with a Michelson interferometer consisting of two radio antennas with spacings of some tens of meters up to 240 meters. They showed that the radio radiation was smaller than 10 arc minutes in size and also detected circular polarization in the Type I bursts. Two other groups had also detected circular polarization at about the same time (David Martyn in Australia and Edward Appleton with James Stanley Hey in the UK).

Modern Radio interferometers consist of widely separated radio telescopes observing the same object that are connected together using coaxial cable, waveguide, optical fiber, or other type of transmission line. This not only increases the total signal collected, it can also be used in a process called Aperture synthesis to vastly increase resolution. This technique works by superposing ("interfering") the signal waves from the different telescopes on the principle that waves that coincide with the same phase will add to each other while two waves that have opposite phases will cancel each other out. This creates a combined telescope that is the size of the antennas furthest apart in the array. In order to produce a high quality image, a large number of different separations between different telescopes are required (the projected separation between any two telescopes as seen from the radio source is called a "baseline") - as many different baselines as possible are required in order to get a good quality image. For example, the Very Large Array has 27 telescopes giving 351 independent baselines at once.

Very Long Baseline Interferometry

Beginning in the 1970s, improvements in the stability of radio telescope receivers permitted telescopes from all over the world (and even in Earth orbit) to be combined to perform Very Long Baseline Interferometry. Instead of physically connecting the antennas, data received at each antenna is paired with timing information, usually from a local atomic clock, and then stored for later analysis on magnetic tape or hard disk. At that later time, the data is correlated with data from other antennas similarly recorded, to produce the resulting image. Using this method it is possible to synthesise an antenna that is effectively the size of the Earth. The large distances between the telescopes enable very high angular resolutions to be achieved, much greater in fact than in any other field of astronomy. At the highest frequencies, synthesised beams less than 1 milliarcsecond are possible.

The Mount Pleasant Radio Telescope is the southern most antenna used in Australia's VLBI network

The pre-eminent VLBI arrays operating today are the Very Long Baseline Array (with telescopes located across North America) and the European VLBI Network (telescopes in Europe, China, South Africa and Puerto Rico). Each array usually operates separately, but occasional projects are observed together producing increased sensitivity. This is referred to as Global VLBI. There are also a VLBI networks, operating in Australia and New Zealand called the LBA (Long Baseline Array), and arrays in Japan, China and South Korea which observe together to form the East-Asian VLBI Network (EAVN).

Since its inception, recording data onto hard media was the only way to bring the data recorded at each telescope together for later correlation. However, the availability today of worldwide, high-bandwidth networks makes it possible to do VLBI in real time. This technique (referred to as e-VLBI) was originally pioneered in Japan, and more recently adopted in Australia and in Europe by the EVN (European VLBI Network) who perform an increasing number of scientific e-VLBI projects per year.

Astronomical Sources

Radio astronomy has led to substantial increases in astronomical knowledge, particularly with the discovery of several classes of new objects, including pulsars, quasars and radio galaxies. This is because radio astronomy allows us to see things that are not detectable in optical astronomy. Such objects represent some of the most extreme and energetic physical processes in the universe.

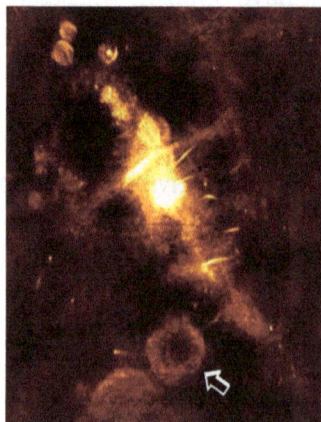

A radio image of the central region of the Milky Way galaxy. The arrow indicates a supernova remnant which is the location of a newly discovered transient, bursting low-frequency radio source GCRT J1745-3009.

The cosmic microwave background radiation was also first detected using radio telescopes. However, radio telescopes have also been used to investigate objects much closer to home, including observations of the Sun and solar activity, and radar mapping of the planets.

High-Energy Astronomy

High energy astronomy is the study of astronomical objects that release electromagnetic radiation of highly energetic wavelengths. It includes X-ray astronomy, gamma-ray astronomy, and extreme UV astronomy, as well as studies of neutrinos and cosmic rays. The physical study of these phenomena is referred to as high-energy astrophysics.

Astronomical objects commonly studied in this field may include black holes, neutron stars, active galactic nuclei, supernovae, supernova remnants, and gamma ray bursts.

Missions

Some space and ground based telescopes that have studied high energy astronomy include the following:

- AGILE (spacecraft)
- AMS-02
- AUGER
- Chandra X-ray Observatory
- Fermi
- H.E.S.S.
- IceCube
- INTEGRAL
- MAGIC
- NuSTAR
- Suzaku (ASTRO-E)
- Swift
- TA
- XMM-Newton - X-ray Multi-Mirror Mission - Newton

Sub-Field of High-Energy Astronomy

X-ray Astronomy

X-ray astronomy is an observational branch of astronomy which deals with the study of X-ray observation and detection from astronomical objects. X-radiation is absorbed by the Earth's atmosphere, so instruments to detect X-rays must be taken to high altitude by balloons, sounding rockets, and satellites. X-ray astronomy is the space science related to a type of space telescope that can see farther than standard light-absorption telescopes, such as the Mauna Kea Observatories, via x-ray radiation.

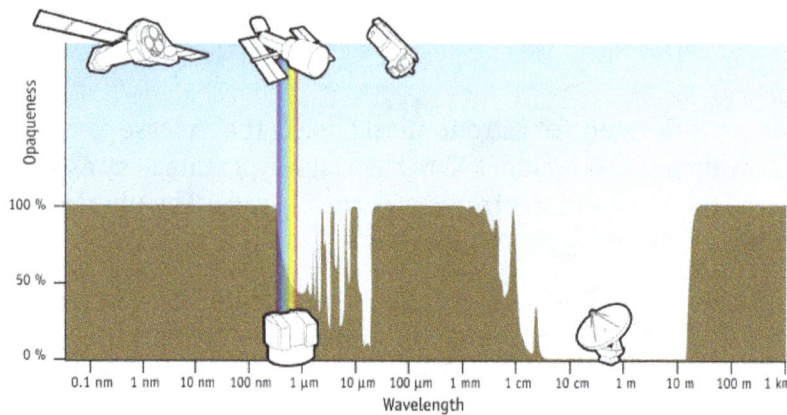

X-rays start at ~0.008 nm and extend across the electromagnetic spectrum to ~8 nm, over which the Earth's atmosphere is opaque.

X-ray emission is expected from astronomical objects that contain extremely hot gasses at temperatures from about a million kelvin (K) to hundreds of millions of kelvin (MK). Although X-rays have been observed emanating from the Sun since the 1940s, the discovery in 1962 of the first cosmic X-ray source was a surprise. This source is called Scorpius X-1 (Sco X-1), the first X-ray source found in the constellation Scorpius. The X-ray emission of Scorpius X-1 is 10,000 times greater than its visual emission, whereas that of the Sun is about a million times less. In addition, the energy output in X-rays is 100,000 times greater than the total emission of the Sun in all wavelengths. Based on discoveries in this new field of X-ray astronomy, starting with Scorpius X-1, Riccardo Giacconi received the Nobel Prize in Physics in 2002. It is now known that such X-ray sources as Sco X-1 are compact stars, such as neutron stars or black holes. Material falling into a black hole may emit X-rays, but the black hole itself does not. The energy source for the X-ray emission is gravity. Infalling gas and dust is heated by the strong gravitational fields of these and other celestial objects.

Many thousands of X-ray sources are known. In addition, the space between galaxies in galaxy clusters is filled with a very hot, but very dilute gas at a temperature between 10 and 100 megakelvins (MK). The total amount of hot gas is five to ten times the total mass in the visible galaxies.

Sounding Rocket Flights

The first sounding rocket flights for X-ray research were accomplished at the White Sands Missile Range in New Mexico with a V-2 rocket on January 28, 1949. A detector was placed in the nose cone section and the rocket was launched in a suborbital flight to an altitude just above the atmosphere.

X-rays from the Sun were detected by the U.S. Naval Research Laboratory Blossom experiment on board. An Aerobee 150 rocket was launched on June 12, 1962 and it detected the first X-rays from other celestial sources (Scorpius X-1).

The largest drawback to rocket flights is their very short duration (just a few minutes above the atmosphere before the rocket falls back to Earth) and their limited field of view. A rocket launched from the United States will not be able to see sources in the southern sky; a rocket launched from Australia will not be able to see sources in the northern sky.

X-ray Quantum Calorimeter (XQC) Project

A launch of the Black Brant 8 Microcalorimeter (XQC-2) at the turn of the century is a part of the joint undertaking by the University of Wisconsin-Madison and NASA's Goddard Space Flight Center known as the X-ray Quantum Calorimeter (XQC) project.

In astronomy, the interstellar medium (or ISM) is the gas and cosmic dust that pervade interstellar space: the matter that exists between the star systems within a galaxy. It fills interstellar space and blends smoothly into the surrounding intergalactic medium. The interstellar medium consists of an extremely dilute (by terrestrial standards) mixture of ions, atoms, molecules, larger dust grains, cosmic rays, and (galactic) magnetic fields. The energy that occupies the same volume, in the form of electromagnetic radiation, is the interstellar radiation field.

Of interest is the hot ionized medium (HIM) consisting of a coronal cloud ejection from star surfaces at 10^6-10^7 K which emits X-rays. The ISM is turbulent and full of structure on all spatial scales. Stars are born deep inside large complexes of molecular clouds, typically a few parsecs in size. During their lives and deaths, stars interact physically with the ISM. Stellar winds from young clusters of stars (often with giant or supergiant HII regions surrounding them) and shock waves created by supernovae inject enormous amounts of energy into their surroundings, which leads to hypersonic turbulence. The resultant structures are stellar wind bubbles and superbubbles of hot gas. The Sun is currently traveling through the Local Interstellar Cloud, a denser region in the low-density Local Bubble.

To measure the spectrum of the diffuse X-ray emission from the interstellar medium over the energy range 0.07 to 1 keV, NASA launched a Black Brant 9 from White Sands Missile Range, New Mexico on May 1, 2008. The Principal Investigator for the mission is Dr. Dan McCammon of the University of Wisconsin.

Balloons

Balloon flights can carry instruments to altitudes of up to 40 km above sea level, where they are above as much as 99.997% of the Earth's atmosphere. Unlike a rocket where data are collected during a brief few minutes, balloons are able to stay aloft for much longer. However, even at such altitudes, much of the X-ray spectrum is still absorbed. X-rays with energies less than 35 keV (5,600 aJ) cannot reach balloons. On July 21, 1964, the Crab Nebula supernova remnant was discovered to be a hard X-ray (15 – 60 keV) source by a scintillation counter flown on a balloon launched from Palestine, Texas, USA. This was likely the first balloon-based detection of X-rays from a discrete cosmic X-ray source.

High-Energy Focusing Telescope

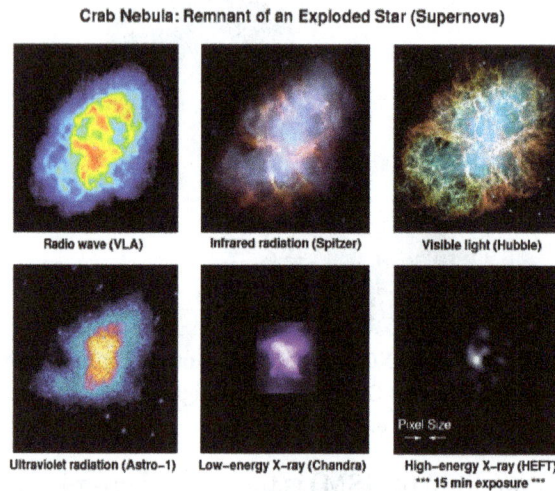

The Crab Nebula is a remnant of an exploded star. This image shows the Crab Nebula in various energy bands, including a hard X-ray image from the HEFT data taken during its 2005 observation run. Each image is 6' wide.

The high-energy focusing telescope (HEFT) is a balloon-borne experiment to image astrophysical sources in the hard X-ray (20–100 keV) band. Its maiden flight took place in May 2005 from Fort Sumner, New Mexico, USA. The angular resolution of HEFT is ~1.5'. Rather than using a grazing-angle X-ray telescope, HEFT makes use of a novel tungsten-silicon multilayer coatings to extend the reflectivity of nested grazing-incidence mirrors beyond 10 keV. HEFT has an energy resolution of 1.0 keV full width at half maximum at 60 keV. HEFT was launched for a 25-hour balloon flight in May 2005. The instrument performed within specification and observed Tau X-1, the Crab Nebula.

High-Resolution Gamma-ray and Hard X-ray Spectrometer (HIREGS)

A balloon-borne experiments called the High-resolution gamma-ray and hard X-ray spectrometer (HIREGS) made observed in X-ray and gamma-rays It was launched from McMurdo Station, Antarctica in December 1991. Steady winds carried the balloon on a circumpolar flight lasting about two weeks.

Rockoons

The rockoon (a portmanteau of rocket and balloon) was a solid fuel rocket that, rather than being immediately lit while on the ground, was first carried into the upper atmosphere by a gas-filled balloon. Then, once separated from the balloon at its maximum height, the rocket was automatically ignited. This achieved a higher altitude, since the rocket did not have to move through the lower thicker air layers that would have required much more chemical fuel.

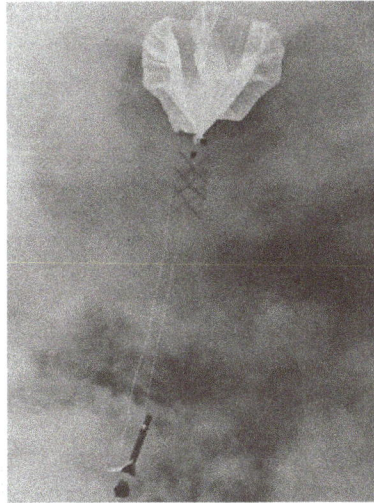

Navy Deacon rockoon photographed just after a shipboard launch in July 1956.

The original concept of "rockoons" was developed by Cmdr. Lee Lewis, Cmdr. G. Halvorson, S. F. Singer, and James A. Van Allen during the Aerobee rocket firing cruise of the USS *Norton Sound* on March 1, 1949.

From July 17 to July 27, 1956, the Naval Research Laboratory (NRL) shipboard launched eight Deacon rockoons for solar ultraviolet and X-ray observations at ~30° N ~121.6° W, southwest of San Clemente Island, apogee: 120 km.

X-ray Astronomy Satellites

X-ray astronomy satellites study X-ray emissions from celestial objects. Satellites, which can detect and transmit data about the X-ray emissions are deployed as part of branch of space science known as X-ray astronomy. Satellites are needed because X-radiation is absorbed by the Earth's atmosphere, so instruments to detect X-rays must be taken to high altitude by balloons, sounding rockets, and satellites.

X-ray Telescopes and Mirrors

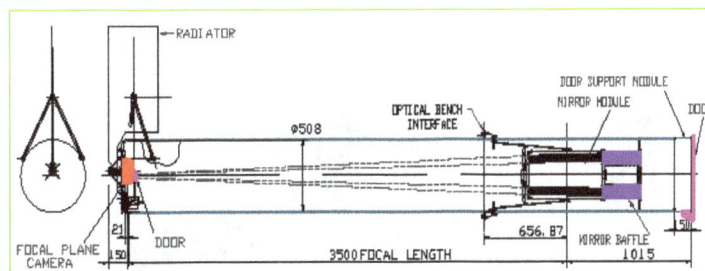

The Swift Gamma-Ray Burst Mission contains a grazing incidence Wolter I telescope (XRT) to focus X-rays onto a state-of-the-art CCD.

X-ray telescopes (XRTs) have varying directionality or imaging ability based on glancing angle reflection rather than refraction or large deviation reflection. This limits them to much narrower fields of view than visible or UV telescopes. The mirrors can be made of ceramic or metal foil.

The first X-ray telescope in astronomy was used to observe the Sun. The first X-ray picture (taken with a grazing incidence telescope) of the Sun was taken in 1963, by a rocket-borne telescope. On April 19, 1960 the very first X-ray image of the sun was taken using a pinhole camera on an Aerobee-Hi rocket.

The utilization of X-ray mirrors for extrasolar X-ray astronomy simultaneously requires:

- the ability to determine the location at the arrival of an X-ray photon in two dimensions and

- a reasonable detection efficiency.

X-ray Astronomy Detectors

Proportional Counter Array on the Rossi X-ray Timing Explorer (RXTE) satellite.

X-ray astronomy detectors have been designed and configured primarily for energy and occasionally for wavelength detection using a variety of techniques usually limited to the technology of the time.

X-ray detectors collect individual X-rays (photons of X-ray electromagnetic radiation) and count the number of photons collected (intensity), the energy (0.12 to 120 keV) of the photons collected, wavelength (~0.008 to 8 nm), or how fast the photons are detected (counts per hour), to tell us about the object that is emitting them.

Astrophysical Sources of X-rays

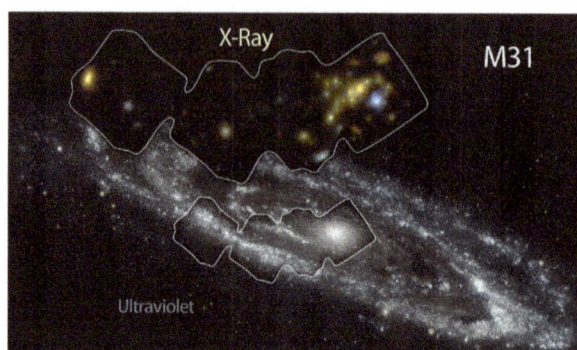

Andromeda Galaxy - in high-energy X-ray and ultraviolet light (released 5 January 2016).

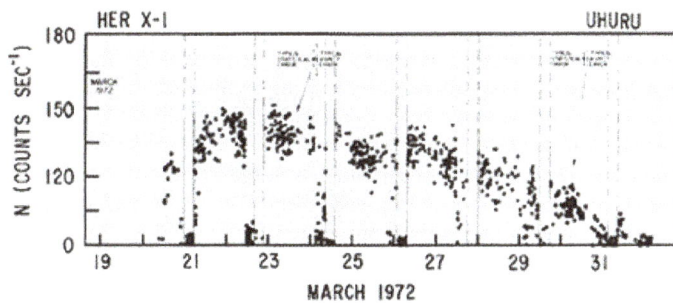

This light curve of Her X-1 shows long term and medium term variability. Each pair of vertical lines delineate the eclipse of the compact object behind its companion star. In this case, the companion is a two solar-mass star with a radius of nearly four times that of our Sun. This eclipse shows us the orbital period of the system, 1.7 days.

Several types of astrophysical objects emit, fluoresce, or reflect X-rays, from galaxy clusters, through black holes in active galactic nuclei (AGN) to galactic objects such as supernova remnants, stars, and binary stars containing a white dwarf (cataclysmic variable stars and super soft X-ray sources), neutron star or black hole (X-ray binaries). Some solar system bodies emit X-rays, the most notable being the Moon, although most of the X-ray brightness of the Moon arises from reflected solar X-rays. A combination of many unresolved X-ray sources is thought to produce the observed X-ray background. The X-ray continuum can arise from bremsstrahlung, black-body radiation, synchrotron radiation, or what is called inverse Compton scattering of lower-energy photons by relativistic electrons, knock-on collisions of fast protons with atomic electrons, and atomic recombination, with or without additional electron transitions.

An intermediate-mass X-ray binary (IMXB) is a binary star system where one of the components is a neutron star or a black hole. The other component is an intermediate mass star.

Hercules X-1 is composed of a neutron star accreting matter from a normal star (HZ Herculis) probably due to Roche lobe overflow. X-1 is the prototype for the massive X-ray binaries although it falls on the borderline, $\sim 2\ M_\odot$, between high- and low-mass X-ray binaries.

Celestial X-ray Sources

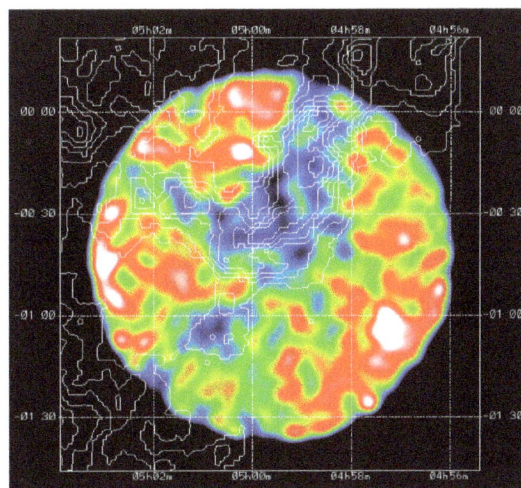

This ROSAT PSPC false-color image is of a portion of a nearby stellar wind superbubble (the Orion-Eridanus Superbubble) stretching across Eridanus and Orion.

The celestial sphere has been divided into 88 constellations. The International Astronomical Union (IAU) constellations are areas of the sky. Each of these contains remarkable X-ray sources. Some of them are have been identified from astrophysical modeling to be galaxies or black holes at the centers of galaxies. Some are pulsars. As with sources already successfully modeled by X-ray astrophysics, striving to understand the generation of X-rays by the apparent source helps to understand the Sun, the universe as a whole, and how these affect us on Earth. Constellations are an astronomical device for handling observation and precision independent of current physical theory or interpretation. Astronomy has been around for a long time. Physical theory changes with time. With respect to celestial X-ray sources, X-ray astrophysics tends to focus on the physical reason for X-ray brightness, whereas X-ray astronomy tends to focus on their classification, order of discovery, variability, resolvability, and their relationship with nearby sources in other constellations.

Within the constellations Orion and Eridanus and stretching across them is a soft X-ray "hot spot" known as the Orion-Eridanus Superbubble, the Eridanus Soft X-ray Enhancement, or simply the Eridanus Bubble, a 25° area of interlocking arcs of Hα emitting filaments. Soft X-rays are emitted by hot gas (T ~ 2–3 MK) in the interior of the superbubble. This bright object forms the background for the "shadow" of a filament of gas and dust. The filament is shown by the overlaid contours, which represent 100 micrometre emission from dust at a temperature of about 30 K as measured by IRAS. Here the filament absorbs soft X-rays between 100 and 300 eV, indicating that the hot gas is located behind the filament. This filament may be part of a shell of neutral gas that surrounds the hot bubble. Its interior is energized by ultraviolet (UV) light and stellar winds from hot stars in the Orion OB1 association. These stars energize a superbubble about 1200 lys across which is observed in the visual (Hα) and X-ray portions of the spectrum.

Proposed (Future) X-ray Observatory Satellites

There are several projects that are proposed for X-ray observatory satellites.

Explorational X-ray Astronomy

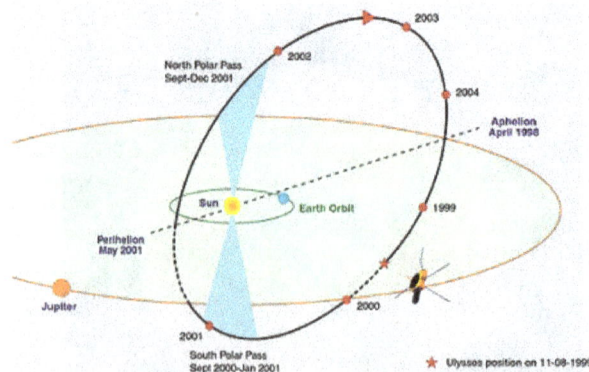

Ulysses' second orbit: it arrived at Jupiter on February 8, 1992, for a swing-by maneuver that increased its inclination to the ecliptic by 80.2 degrees.

Usually observational astronomy is considered to occur on Earth's surface (or beneath it in neutrino astronomy). The idea of limiting observation to Earth includes orbiting the Earth. As soon as the observer leaves the cozy confines of Earth, the observer becomes a deep space explorer. Except for Explorer 1 and Explorer 3 and the earlier satellites in the series, usually if a probe is going to be a deep space explorer it leaves the Earth or an orbit around the Earth.

For a satellite or space probe to qualify as a deep space X-ray astronomer/explorer or "astronobot"/explorer, all it needs to carry aboard is an XRT or X-ray detector and leave Earth orbit.

Ulysses is launched October 6, 1990, and reached Jupiter for its "gravitational slingshot" in February 1992. It passed the south solar pole in June 1994 and crossed the ecliptic equator in February 1995. The solar X-ray and cosmic gamma-ray burst experiment (GRB) had 3 main objectives: study and monitor solar flares, detect and localize cosmic gamma-ray bursts, and in-situ detection of Jovian aurorae. Ulysses was the first satellite carrying a gamma burst detector which went outside the orbit of Mars. The hard X-ray detectors operated in the range 15–150 keV. The detectors consisted of 23-mm thick × 51-mm diameter CsI(Tl) crystals mounted via plastic light tubes to photomultipliers. The hard detector changed its operating mode depending on (1) measured count rate, (2) ground command, or (3) change in spacecraft telemetry mode. The trigger level was generally set for 8-sigma above background and the sensitivity is 10^{-6} erg/cm^2 (1 nJ/m^2). When a burst trigger is recorded, the instrument switches to record high resolution data, recording it to a 32-kbit memory for a slow telemetry read out. Burst data consist of either 16 s of 8-ms resolution count rates or 64 s of 32-ms count rates from the sum of the 2 detectors. There were also 16 channel energy spectra from the sum of the 2 detectors (taken either in 1, 2, 4, 16, or 32 second integrations). During 'wait' mode, the data were taken either in 0.25 or 0.5 s integrations and 4 energy channels (with shortest integration time being 8 s). Again, the outputs of the 2 detectors were summed.

The Ulysses soft X-ray detectors consisted of 2.5-mm thick × 0.5 cm^2 area Si surface barrier detectors. A 100 mg/cm^2 beryllium foil front window rejected the low energy X-rays and defined a conical FOV of 75° (half-angle). These detectors were passively cooled and operate in the temperature range −35 to −55 °C. This detector had 6 energy channels, covering the range 5–20 keV.

X-Rays from Pluto

Theoretical X-ray Astronomy

Theoretical X-ray astronomy is a branch of theoretical astronomy that deals with the theoretical astrophysics and theoretical astrochemistry of X-ray generation, emission, and detection as applied to astronomical objects.

Like theoretical astrophysics, theoretical X-ray astronomy uses a wide variety of tools which include analytical models to approximate the behavior of a possible X-ray source and computational numerical simulations to approximate the observational data. Once potential observational consequences are available they can be compared with experimental observations. Observers can look for data that refutes a model or helps in choosing between several alternate or conflicting models.

Theorists also try to generate or modify models to take into account new data. In the case of an inconsistency, the general tendency is to try to make minimal modifications to the model to fit the data. In some cases, a large amount of inconsistent data over time may lead to total abandonment of a model.

Most of the topics in astrophysics, astrochemistry, astrometry, and other fields that are branches of astronomy studied by theoreticians involve X-rays and X-ray sources. Many of the beginnings for a theory can be found in an Earth-based laboratory where an X-ray source is built and studied.

Dynamos

Dynamo theory describes the process through which a rotating, convecting, and electrically conducting fluid acts to maintain a magnetic field. This theory is used to explain the presence of anomalously long-lived magnetic fields in astrophysical bodies. If some of the stellar magnetic fields are really induced by dynamos, then field strength might be associated with rotation rate.

Astronomical Models

From the observed X-ray spectrum, combined with spectral emission results for other wavelength ranges, an astronomical model addressing the likely source of X-ray emission can be constructed. For example, with Scorpius X-1 the X-ray spectrum steeply drops off as X-ray energy increases up to 20 keV, which is likely for a thermal-plasma mechanism. In addition, there is no radio emission, and the visible continuum is roughly what would be expected from a hot plasma fitting the observed X-ray flux. The plasma could be a coronal cloud of a central object or a transient plasma, where the energy source is unknown, but could be related to the idea of a close binary.

Images released to celebrate the International Year of Light 2015(Chandra X-Ray Observatory).

In the Crab Nebula X-ray spectrum there are three features that differ greatly from Scorpius X-1: its spectrum is much harder, its source diameter is in light-years (ly)s, not astronomical units (AU), and its radio and optical synchrotron emission are strong. Its overall X-ray luminosity rivals the optical emission and could be that of a nonthermal plasma. However, the Crab Nebula appears as an X-ray source that is a central freely expanding ball of dilute plasma, where the energy content is 100 times the total energy content of the large visible and radio portion, obtained from the unknown source.

The "Dividing Line" as giant stars evolve to become red giants also coincides with the Wind and Coronal Dividing Lines. To explain the drop in X-ray emission across these dividing lines, a number of models have been proposed:

- low transition region densities, leading to low emission in coronae,

- high-density wind extinction of coronal emission,

- only cool coronal loops become stable,

- changes in a magnetic field structure to that an open topology, leading to a decrease of magnetically confined plasma, or

- changes in the magnetic dynamo character, leading to the disappearance of stellar fields leaving only small-scale, turbulence-generated fields among red giants.

Analytical X-ray Astronomy

Analytical X-ray astronomy is applied to an astronomy puzzle in an attempt to provide an acceptable solution. Consider the following puzzle.

High-mass X-ray binaries (HMXBs) are composed of OB supergiant companion stars and compact objects, usually neutron stars (NS) or black holes (BH). Supergiant X-ray binaries (SGXBs) are HMXBs in which the compact objects orbit massive companions with orbital periods of a few days (3–15 d), and in circular (or slightly eccentric) orbits. SGXBs show typical the hard X-ray spectra of accreting pulsars and most show strong absorption as obscured HMXBs. X-ray luminosity (L_x) increases up to 10^{36} erg·s^{-1} (10^{29} watts).

The mechanism triggering the different temporal behavior observed between the classical SGXBs and the recently discovered supergiant fast X-ray transients (SFXT)s is still debated.

Aim: use the discovery of long orbits (>15 d) to help discriminate between emission models and perhaps bring constraints on the models.

Method: analyze archival data on various SGXBs such as has been obtained by INTEGRAL for candidates exhibiting long orbits. Build short- and long-term light curves. Perform a timing analysis in order to study the temporal behavior of each candidate on different time scales.

Compare various astronomical models:

- direct spherical accretion

- Roche-Lobe overflow via an accretion disk on the compact object.

Draw some conclusions: for example, the SGXB SAX J1818.6-1703 was discovered by BeppoSAX in 1998, identified as a SGXB of spectral type between O9I–B1I, which also displayed short and bright flares and an unusually very low quiescent level leading to its classification as a SFXT. The analysis indicated an unusually long orbital period: 30.0 ± 0.2 d and an elapsed accretion phase of ~6 d implying an elliptical orbit and possible supergiant spectral type between B0.5-1I with eccentricities $e \sim 0.3$–0.4. The large variations in the X-ray flux can be explained through accretion of macro-clumps formed within the stellar wind.

Choose which model seems to work best: for SAX J1818.6-1703 the analysis best fits the model that predicts SFXTs behave as SGXBs with different orbital parameters; hence, different temporal behavior.

Stellar X-ray Astronomy

Stellar X-ray astronomy is said to have started on April 5, 1974, with the detection of X-rays from Capella. A rocket flight on that date briefly calibrated its attitude control system when a star sensor pointed the payload axis at Capella (α Aur). During this period, X-rays in the range 0.2–1.6 keV were detected by an X-ray reflector system co-aligned with the star sensor. The X-ray luminosity of $L_x = 10^{31}$ erg·s^{-1} (10^{24} W) is four orders of magnitude above the Sun's X-ray luminosity.

Eta Carinae

New X-ray observations by the Chandra X-ray Observatory show three distinct structures: an outer, horseshoe-shaped ring about 2 light years in diameter, a hot inner core about 3 light-months in diameter, and a hot central source less than 1 light-month in diameter which may contain the superstar that drives the whole show. The outer ring provides evidence of another large explosion that occurred over 1,000 years ago. These three structures around Eta Carinae are thought to represent shock waves produced by matter rushing away from the superstar at supersonic speeds. The temperature of the shock-heated gas ranges from 60 MK in the central regions to 3 MK on the horseshoe-shaped outer structure. "The Chandra image contains some puzzles for existing ideas of how a star can produce such hot and intense X-rays," says Prof. Kris Davidson of the University of Minnesota. Davidson is principal investigator for the Eta Carina observations by the Hubble Space telescope. "In the most popular theory, X-rays are made by colliding gas streams from two stars so close together that they'd look like a point source to us. But what happens to gas streams that escape to farther distances? The extended hot stuff in the middle of the new image gives demanding new conditions for any theory to meet."

Classified as a Peculiar star, Eta Carinae exhibits a superstar at its center as seen in this image from Chandra X-ray Observatory. Credit: Chandra Science Center and NASA.

Stellar Coronae

Coronal stars, or stars within a coronal cloud, are ubiquitous among the stars in the cool half of the Hertzsprung-Russell diagram. Experiments with instruments aboard Skylab and Copernicus have been used to search for soft X-ray emission in the energy range ~0.14–0.284 keV from stellar coronae. The experiments aboard ANS succeeded in finding X-ray signals from Capella and Sirius (α CMa). X-ray emission from an enhanced solar-like corona was proposed for the first time. The high temperature of Capella's corona as obtained from the first coronal X-ray spectrum of Capella using HEAO 1 required magnetic confinement unless it was a free-flowing coronal wind.

In 1977 Proxima Centauri is discovered to be emitting high-energy radiation in the XUV. In 1978, α Cen was identified as a low-activity coronal source. With the operation of the Einstein observatory, X-ray emission was recognized as a characteristic feature common to a wide range of stars covering essentially the whole Hertzsprung-Russell diagram. The Einstein initial survey led to significant insights:

- X-ray sources abound among all types of stars, across the Hertzsprung-Russell diagram and across most stages of evolution,

- the X-ray luminosities and their distribution along the main sequence were not in agreement with the long-favored acoustic heating theories, but were now interpreted as the effect of magnetic coronal heating, and

- stars that are otherwise similar reveal large differences in their X-ray output if their rotation period is different.

To fit the medium-resolution spectrum of UX Ari, subsolar abundances were required.

Stellar X-ray astronomy is contributing toward a deeper understanding of

- magnetic fields in magnetohydrodynamic dynamos,

- the release of energy in tenuous astrophysical plasmas through various plasma-physical processes, and

- the interactions of high-energy radiation with the stellar environment.

Current wisdom has it that the massive coronal main sequence stars are late-A or early F stars, a conjecture that is supported both by observation and by theory.

Unstable winds

Given the lack of a significant outer convection zone, theory predicts the absence of a magnetic dynamo in earlier A stars. In early stars of spectral type O and B, shocks developing in unstable winds are the likely source of X-rays.

Coolest M Dwarfs

Beyond spectral type M5, the classical $\alpha\omega$ dynamo can no longer operate as the internal structure of dwarf stars changes significantly: they become fully convective. As a distributed (or α^2) dynamo may become relevant, both the magnetic flux on the surface and the topology of the magnetic fields in the corona should systematically change across this transition, perhaps resulting in some discontinuities in the X-ray characteristics around spectral class dM5. However, observations do not seem to support this picture: long-time lowest-mass X-ray detection, VB 8 (M7e V), has shown steady emission at levels of X-ray luminosity (L_x) $\approx 10^{26}$ erg·s^{-1} (10^{19} W) and flares up to an order of magnitude higher. Comparison with other late M dwarfs shows a rather continuous trend.

Strong X-ray Emission from Herbig Ae/Be Stars

Herbig Ae/Be stars are pre-main sequence stars. As to their X-ray emission properties, some are

- reminiscent of hot stars,
- others point to coronal activity as in cool stars, in particular the presence of flares and very high temperatures.

The nature of these strong emissions has remained controversial with models including

- unstable stellar winds,
- colliding winds,
- magnetic coronae,
- disk coronae,
- wind-fed magnetospheres,
- accretion shocks,
- the operation of a shear dynamo,
- the presence of unknown late-type companions.

K Giants

The FK Com stars are giants of spectral type K with an unusually rapid rotation and signs of extreme activity. Their X-ray coronae are among the most luminous ($L_x \geq 10^{32}$ erg·s^{-1} or 10^{25} W) and the hottest known with dominant temperatures up to 40 MK. However, the current popular hypothesis involves a merger of a close binary system in which the orbital angular momentum of the companion is transferred to the primary.

Pollux is the brightest star in the constellation Gemini, despite its Beta designation, and the 17th brightest in the sky. Pollux is a giant orange K star that makes an interesting color contrast with its white "twin", Castor. Evidence has been found for a hot, outer, magnetically supported corona around Pollux, and the star is known to be an X-ray emitter.

Amateur X-ray Astronomy

Collectively, amateur astronomers observe a variety of celestial objects and phenomena sometimes with equipment that they build themselves. The United States Air Force Academy (US-AFA) is the home of the US's only undergraduate satellite program, and has and continues to develop the FalconLaunch sounding rockets. In addition to any direct amateur efforts to put

X-ray astronomy payloads into space, there are opportunities that allow student-developed experimental payloads to be put on board commercial sounding rockets as a free-of-charge ride.

There are major limitations to amateurs observing and reporting experiments in X-ray astronomy: the cost of building an amateur rocket or balloon to place a detector high enough and the cost of appropriate parts to build a suitable X-ray detector.

History of X-ray Astronomy

In 1927, E.O. Hulburt of the US Naval Research Laboratory and associates Gregory Breit and Merle A. Tuve of the Carnegie Institution of Washington explored the possibility of equipping Robert H. Goddard's rockets to explore the upper atmosphere. "Two years later, he proposed an experimental program in which a rocket might be instrumented to explore the upper atmosphere, including detection of ultraviolet radiation and X-rays at high altitudes".

NRL scientists J. D. Purcell, C. Y. Johnson, and Dr. F. S. Johnson are among those recovering instruments from a V-2 used for upper atmospheric research above the New Mexico desert. This is V-2 number 54, launched January 18, 1951, (photo by Dr. Richard Tousey, NRL).

In the late 1930s, the presence of a very hot, tenuous gas surrounding the Sun was inferred indirectly from optical coronal lines of highly ionized species. The Sun has been known to be surrounded by a hot tenuous corona. In the mid-1940s radio observations revealed a radio corona around the Sun.

The beginning of the search for X-ray sources from above the Earth's atmosphere was on August 5, 1948 12:07 GMT. A US Army (formerly German) V-2 rocket as part of Project Hermes was launched from White Sands Proving Grounds. The first solar X-rays were recorded by T. Burnight.

Through the 1960s, 70s, 80s, and 90s, the sensitivity of detectors increased greatly during the 60 years of X-ray astronomy. In addition, the ability to focus X-rays has developed enormously—allowing the production of high-quality images of many fascinating celestial objects.

Major Questions in X-ray Astronomy

As X-ray astronomy uses a major spectral probe to peer into source, it is a valuable tool in efforts to understand many puzzles.

Stellar Magnetic Fields

Magnetic fields are ubiquitous among stars, yet we do not understand precisely why, nor have we fully understood the bewildering variety of plasma physical mechanisms that act in stellar environments. Some stars, for example, seem to have magnetic fields, fossil stellar magnetic fields left over from their period of formation, while others seem to generate the field anew frequently.

Extrasolar X-ray Source Astrometry

With the initial detection of an extrasolar X-ray source, the first question usually asked is "What is the source?" An extensive search is often made in other wavelengths such as visible or radio for possible coincident objects. Many of the verified X-ray locations still do not have readily discernible sources. X-ray astrometry becomes a serious concern that results in ever greater demands for finer angular resolution and spectral radiance.

There are inherent difficulties in making X-ray/optical, X-ray/radio, and X-ray/X-ray identifications based solely on positional coincidents, especially with handicaps in making identifications, such as the large uncertainties in positional determinants made from balloons and rockets, poor source separation in the crowded region toward the galactic center, source variability, and the multiplicity of source nomenclature.

X-ray source counterparts to stars can be identified by calculating the angular separation between source centroids and position of the star. The maximum allowable separation is a compromise between a larger value to identify as many real matches as possible and a smaller value to minimize the probability of spurious matches. "An adopted matching criterion of 40" finds nearly all possible X-ray source matches while keeping the probability of any spurious matches in the sample to 3%."

Solar X-ray Astronomy

All of the detected X-ray sources at, around, or near the Sun are within or associated with the coronal cloud which is its outer atmosphere.

Coronal Heating Problem

In the area of solar X-ray astronomy, there is the coronal heating problem. The photosphere of the Sun has an effective temperature of 5,570 K yet its corona has an average temperature of $1-2 \times 10^6$ K. However, the hottest regions are $8-20 \times 10^6$ K. The high temperature of the corona shows that it is heated by something other than direct heat conduction from the photosphere.

It is thought that the energy necessary to heat the corona is provided by turbulent motion in the convection zone below the photosphere, and two main mechanisms have been proposed to explain coronal heating. The first is wave heating, in which sound, gravitational or magnetohydrodynamic waves are produced by turbulence in the convection zone. These waves travel upward and dissipate in the corona, depositing their energy in the ambient gas in the form of heat. The other is magnetic heating, in which magnetic energy is continuously built up by photospheric motion and released through magnetic reconnection in the form of large solar flares and myriad similar but smaller events—nanoflares.

Currently, it is unclear whether waves are an efficient heating mechanism. All waves except Alfvén waves have been found to dissipate or refract before reaching the corona. In addition, Alfvén waves do not easily dissipate in the corona. Current research focus has therefore shifted towards flare heating mechanisms.

Coronal Mass Ejection

A coronal mass ejection (CME) is an ejected plasma consisting primarily of electrons and protons (in addition to small quantities of heavier elements such as helium, oxygen, and iron), plus the entraining coronal closed magnetic field regions. Evolution of these closed magnetic structures in response to various photospheric motions over different time scales (convection, differential rotation, meridional circulation) somehow leads to the CME. Small-scale energetic signatures such as plasma heating (observed as compact soft X-ray brightening) may be indicative of impending CMEs.

The soft X-ray sigmoid (an S-shaped intensity of soft X-rays) is an observational manifestation of the connection between coronal structure and CME production. "Relating the sigmoids at X-ray (and other) wavelengths to magnetic structures and current systems in the solar atmosphere is the key to understanding their relationship to CMEs."

The first detection of a Coronal mass ejection (CME) as such was made on December 1, 1971 by R. Tousey of the US Naval Research Laboratory using OSO 7. Earlier observations of v or even phenomena observed visually during solar eclipses are now understood as essentially the same thing.

The largest geomagnetic perturbation, resulting presumably from a "prehistoric" CME, coincided with the first-observed solar flare, in 1859. The flare was observed visually by Richard Christopher Carrington and the geomagnetic storm was observed with the recording magnetograph at Kew Gardens. The same instrument recorded a crotchet, an instantaneous perturbation of the Earth's ionosphere by ionizing soft X-rays. This could not easily be understood at the time because it predated the discovery of X-rays (by Roentgen) and the recognition of the ionosphere (by Kennelly and Heaviside).

Exotic X-ray Sources

A microquasar is a smaller cousin of a quasar that is a radio emitting X-ray binary, with an often resolvable pair of radio jets. LSI+61°303 is a periodic, radio-emitting binary system that is also the gamma-ray source, CG135+01. Observations are revealing a growing number of recurrent X-ray transients, characterized by short outbursts with very fast rise times (tens of minutes) and typical durations of a few hours that are associated with OB supergiants and hence define a new class of massive X-ray binaries: Supergiant Fast X-ray Transients (SFXTs). Observations made by Chandra indicate the presence of loops and rings in the hot X-ray emitting gas that surrounds Messier 87. A magnetar is a type of neutron star with an extremely powerful magnetic field, the decay of which powers the emission of copious amounts of high-energy electromagnetic radiation, particularly X-rays and gamma rays.

X-ray Dark Stars

During the solar cycle, as shown in the sequence of images at right, at times the Sun is almost X-ray dark, almost an X-ray variable. Betelgeuse, on the other hand, appears to be always X-ray dark. Hardly any X-rays are emitted by red giants. There is a rather abrupt onset of X-ray emission around spectral type A7-F0, with a large range of luminosities developing across spectral class F. Altair is spectral type A7V and Vega is A0V. Altair's total X-ray luminosity is at least an order of magnitude larger than the X-ray luminosity for Vega. The outer convection zone of early F stars is expected to be very shallow and absent in A-type dwarfs, yet the acoustic flux from the interior reaches a maximum for late A and early F stars provoking investigations of magnetic activity in A-type stars along three principal lines. Chemically peculiar stars of spectral type Bp or Ap are appreciable magnetic radio sources, most Bp/Ap stars remain undetected, and of those reported early on as producing X-rays only few of them can be identified as probably single stars. X-ray observations offer the possibility to detect (X-ray dark) planets as they eclipse part of the corona of their parent star while in transit. "Such methods are particularly promising for low-mass stars as a Jupiter-like planet could eclipse a rather significant coronal area."

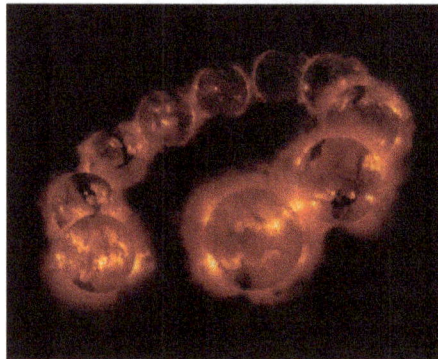

A solar cycle: a montage of ten years' worth of Yohkoh SXT images, demonstrating the variation in solar activity during a sunspot cycle, from after August 30, 1991, at the peak of cycle 22, to September 6, 2001, at the peak of cycle 23. Credit: the Yohkoh mission of Institute of Space and Astronautical Science (ISAS, Japan) and NASA (US).

X-ray Dark Planet/Comet

X-ray observations offer the possibility to detect (X-ray dark) planets as they eclipse part of the corona of their parent star while in transit. "Such methods are particularly promising for low-mass stars as a Jupiter-like planet could eclipse a rather significant coronal area."

As X-ray detectors have become more sensitive, they have observed that some planets and other normally X-ray non-luminescent celestial objects under certain conditions emit, fluoresce, or reflect X-rays.

Comet Lulin

NASA's Swift Gamma-Ray Burst Mission satellite was monitoring Comet Lulin as it closed to 63 Gm of Earth. For the first time, astronomers can see simultaneous UV and X-ray images of a comet. "The solar wind—a fast-moving stream of particles from the sun—interacts with the comet's broader cloud of atoms. This causes the solar wind to light up with X-rays, and that's what Swift's XRT sees", said Stefan Immler, of the Goddard Space Flight Center. This interaction, called charge exchange, results in X-rays from most comets when they pass within about three times Earth's distance from the Sun. Because Lulin is so active, its atomic cloud is especially dense. As a result, the X-ray-emitting region extends far sunward of the comet.

Comet Lulin was passing through the constellation Libra when Swift imaged it on January 28, 2009. This image merges data acquired by Swift's Ultraviolet/Optical Telescope (blue and green) and X-Ray Telescope (red). At the time of the observation, the comet was 99.5 million miles from Earth and 115.3 million miles from the Sun.

Single X-ray Stars

In addition to the Sun there are many unary stars or star systems throughout the galaxy that emit X-rays. β Hydri (G2 IV) is a normal single, post main-sequence subgiant star, T_{eff} = 5800 K. It exhibits coronal X-ray fluxes.

The benefit of studying single stars is that it allows measurements free of any effects of a companion or being a part of a multiple star system. Theories or models can be more readily tested.

Gamma-ray Astronomy

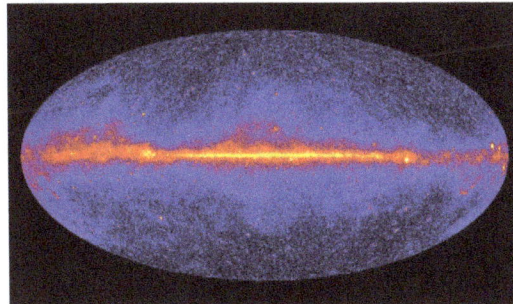

First survey of the sky at energies above 1 GeV, collected by the Fermi Gamma-ray Space Telescope in three years of observation (2009 to 2011)

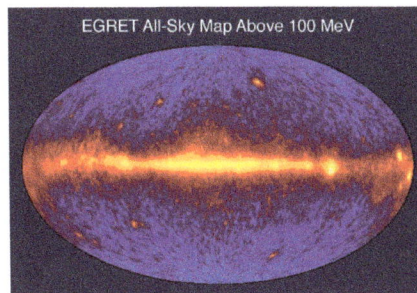

The sky at energies above 100 MeV observed by the Energetic Gamma Ray Experiment Telescope (EGRET) of the Compton Gamma Ray Observatory (CGRO) satellite (1991–2000)

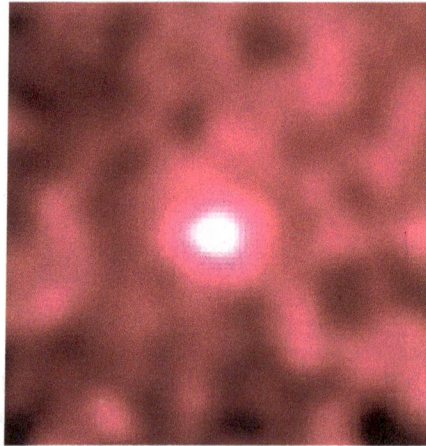

The Moon as seen by the Energetic Gamma Ray Experiment Telescope (EGRET), in gamma rays of greater than 20 MeV. These are produced by cosmic ray bombardment of its surface.

Gamma-ray astronomy is the astronomical observation of gamma rays, the most energetic form of electromagnetic radiation, with photon energies above 100 keV. Radiation below 100 keV is classified as X-rays and is the subject of X-ray astronomy.

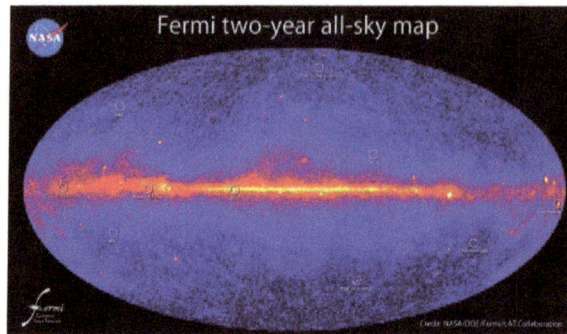

September 02 2011 Fermi Second catalog of Gamma Ray Sources constructed over 2 years. An all sky image showing energies greater than 1 biilion electron volts (1 GeV) ub. Brighter colors indicate gamma-ray sources.

Gamma rays in the MeV range are generated in solar flares (and even in the Earth's atmosphere), but gamma rays in the GeV range do not originate in the Solar System and are important in the study of extrasolar, and especially extra-galactic astronomy. The mechanisms emitting gamma rays are diverse, mostly identical with those emitting X-rays but at higher energies, including electron-positron annihilation, the Inverse Compton Effect, and in some cases also the decay of radioactive material (gamma decay) in space reflecting extreme events such as supernovae and hypernovae, and the behaviour of matter under extreme conditions, as in pulsars and blazars. The highest photon energies measured to date are in the TeV range, the record being held by the Crab Pulsar in 2004, yielding photons with as much as 80 TeV.

Detector Technology

Observation of gamma rays first became possible in the 1960s. Their observation is much more problematic than that of X-rays or of visible light, because gamma-rays are comparatively rare, even a "bright" source needing an observation time of several minutes before it is even detected, and because gamma rays are difficult to focus, resulting in a very low resolution. The most recent generation of gamma-ray telescopes (2000s) have a resolution of the order of 6 arc minutes in the GeV range (seeing the Crab Nebula as a single "pixel"), compared to 0.5 arc seconds seen in the low energy X-ray (1 keV) range by the Chandra X-ray Observatory (1999), and about 1.5 arc minutes in the high energy X-ray (100 keV) range seen by High-Energy Focusing Telescope (2005).

Very energetic gamma rays, with photon energies over ~30 GeV, can also be detected by ground based experiments. The extremely low photon fluxes at such high energies require detector effective areas that are impractically large for current space-based instruments. Fortunately, such high-energy photons produce extensive showers of secondary particles in the atmosphere that can be observed on the ground, both directly by radiation counters and optically via the Cherenkov light which the ultra-relativistic shower particles emit. The Imaging Atmospheric Cherenkov Telescope technique currently achieves the highest sensitivity.

Gamma radiation in the TeV range emanating from the Crab Nebula was first detected in 1989 by the Whipple Observatory at Mt. Hopkins, in Arizona in the USA. Modern Cherenkov telescope experiments like H.E.S.S., VERITAS, MAGIC, and CANGAROO III can detect the Crab Nebula in a few minutes. The most energetic photons (up to 16 TeV) observed from an extragalactic object originate from the blazar, Markarian 501 (Mrk 501). These measurements were done by the High-Energy-Gamma-Ray Astronomy (HEGRA) air Cherenkov telescopes.

Gamma-ray astronomy observations are still limited by non-gamma-ray backgrounds at lower energies, and, at higher energy, by the number of photons that can be detected. Larger area detectors and better background suppression are essential for progress in the field. A discovery in 2012 may allow focusing gamma-ray telescopes. At photon energies greater than 700 keV, the index of refraction starts to increase again.

Early History

Long before experiments could detect gamma rays emitted by cosmic sources, scientists had known that the universe should be producing them. Work by Eugene Feenberg and Henry Primakoff in 1948, Sachio Hayakawa and I.B. Hutchinson in 1952, and, especially, Philip Morrison in 1958 had led scientists to believe that a number of different processes which were occurring in the universe would result in gamma-ray emission. These processes included cosmic ray interactions with interstellar gas, supernova explosions, and interactions of energetic electrons with magnetic fields. However, it was not until the 1960s that our ability to actually detect these emissions came to pass.

Most gamma rays coming from space are absorbed by the Earth's atmosphere, so gamma-ray astronomy could not develop until it was possible to get detectors above all or most of the atmosphere using balloons and spacecraft. The first gamma-ray telescope carried into orbit, on the Explorer 11 satellite in 1961, picked up fewer than 100 cosmic gamma-ray photons. They appeared to come from all directions in the Universe, implying some sort of uniform "gamma-ray background". Such a background would be expected from the interaction of cosmic rays (very energetic charged particles in space) with interstellar gas.

The first true astrophysical gamma-ray sources were solar flares, which revealed the strong 2.223 MeV line predicted by Morrison. This line results from the formation of deuterium via the union of a neutron and proton; in a solar flare the neutrons appear as secondaries from interactions of high-energy ions accelerated in the flare process. These first gamma-ray line observations were from OSO-3, OSO-7, and the Solar Maximum Mission, the latter spacecraft launched in 1980. The solar observations inspired theoretical work by Reuven Ramaty and others.

Significant gamma-ray emission from our galaxy was first detected in 1967 by the detector aboard the OSO-3 satellite. It detected 621 events attributable to cosmic gamma rays. However, the field of gamma-ray astronomy took great leaps forward with the SAS-2 (1972) and the COS-B (1975–1982) satellites. These two satellites provided an exciting view into the high-energy universe (sometimes called the 'violent' universe, because the kinds of events in space that produce gamma rays tend to be high-speed collisions and similar processes). They confirmed the earlier findings of the gamma-ray background, produced the first detailed map of the sky at gamma-ray wavelengths, and detected a number of point sources. However the resolution of the instruments was insufficient to identify most of these point sources with specific visible stars or stellar systems.

A discovery in gamma-ray astronomy came in the late 1960s and early 1970s from a constellation of military defense satellites. Detectors on board the Vela satellite series, designed to detect flashes of gamma rays from nuclear bomb blasts, began to record bursts of gamma rays from deep space rather than the vicinity of the Earth. Later detectors determined that these gamma-ray bursts are seen to last for fractions of a second to minutes, appearing suddenly from unexpected directions, flickering, and then fading after briefly dominating the gamma-ray sky. Studied since the mid-1980s with instruments on board a variety of satellites and space probes, including Soviet Venera spacecraft and the Pioneer Venus Orbiter, the sources of these enigmatic high-energy flashes remain a mystery. They appear to come from far away in the Universe, and currently the most likely theory seems to be that at least some of them come from so-called *hypernova* explosions—supernovas creating black holes rather than neutron stars.

Nuclear gamma rays were observed from the solar flares of August 4 and 7, 1972, and November 22, 1977. A solar flare is an explosion in a solar atmosphere and was originally detected visually in our own sun. Solar flares create massive amounts of radiation across the full electromagnetic spectrum from the longest wavelength, radio waves, to high energy gamma rays. The correlations of the high energy electrons energized during the flare and the gamma rays are mostly caused by nuclear combinations of high energy protons and other heavier ions. These gamma rays can be observed and allow scientists to determine the major results of the energy released, which is not provided by the emissions from other wavelengths.

1980s to 1990s

On June 19, 1988, from Birigüi (50° 20' W 21° 20' S) at 10:15 UTC a balloon launch occurred which carried two NaI(Tl) detectors (600 cm^2 total area) to an air pressure altitude of 5.5 mb for a total observation time of 6 hr. The supernova SN1987A in the Large Magellanic Cloud (LMC) was discovered on February 23, 1987, and its progenitor was a blue supergiant, (Sk -69 202), with luminosity of 2-5 x 10^{38} erg/s. The 847 keV and 1238 keV gamma-ray lines from ^{56}Co decay have been detected.

During its High Energy Astronomy Observatory program in 1977, NASA announced plans to build a "great observatory" for gamma-ray astronomy. The Compton Gamma-Ray Observatory

(CGRO) was designed to take advantage of the major advances in detector technology during the 1980s, and was launched in 1991. The satellite carried four major instruments which have greatly improved the spatial and temporal resolution of gamma-ray observations. The CGRO provided large amounts of data which are being used to improve our understanding of the high-energy processes in our Universe. CGRO was de-orbited in June 2000 as a result of the failure of one of its stabilizing gyroscopes.

BeppoSAX was launched in 1996 and deorbited in 2003. It predominantly studied X-rays, but also observed gamma-ray bursts. By identifying the first non-gamma ray counterparts to gamma-ray bursts, it opened the way for their precise position determination and optical observation of their fading remnants in distant galaxies.

The High Energy Transient Explorer 2 (HETE-2) was launched in October 2000 (on a nominally 2 year mission) and was still operational (but fading) in March 2007.

Recent Observations

Swift, a NASA spacecraft, was launched in 2004 and carries the BAT instrument for gamma-ray burst observations. Following BeppoSAX and HETE-2, it has observed numerous X-ray and optical counterparts to bursts, leading to distance determinations and detailed optical follow-up. These have established that most bursts originate in the explosions of massive stars (supernovas and hypernovas) in distant galaxies. It is still operational in 2015.

Currently the (other) main space-based gamma-ray observatories are the INTErnational Gamma-Ray Astrophysics Laboratory (INTEGRAL), Fermi, and the Astrorivelatore Gamma ad Immagini Leggero (AGILE).

INTEGRAL (launched on 17 October 2002) is an ESA mission with additional contributions from the Czech Republic, Poland, US, and Russia.

AGILE is an all Italian small mission by ASI, INAF and INFN collaboration. It was successfully launched by the Indian PSLV-C8 rocket from the Sriharikota ISRO base on April 23, 2007.

Fermi was launched by NASA on 11 June 2008. It includes LAT, the Large Area Telescope, and GBM, the GLAST Burst Monitor, for studying gamma-ray bursts.

In November 2010, using the Fermi Gamma-ray Space Telescope, two gigantic gamma-ray bubbles, spanning about 25,000 light-years across, were detected at the heart of our galaxy. These bubbles of high-energy radiation are suspected as erupting from a massive black hole or evidence of a burst of star formations from millions of years ago. They were discovered after scientists filtered out the "fog of background gamma-rays suffusing the sky". This discovery confirmed previous clues that a large unknown "structure" was in the center of the Milky Way.

In 2011 the Fermi team released its second catalog of gamma-ray sources detected by the satellite's Large Area Telescope (LAT), which produced an inventory of 1,873 objects shining with the highest-energy form of light. 57% of the sources are Blazars. Over half of the sources are active galaxies, their central black holes created gamma-ray emissions detected by the LAT. One third of the sources have not been detected in other wavelengths.

Two gigantic gamma-ray bubbles at the heart of the Milky Way.

Top 10 Gamma Ray Sources

The Fermi team created a year 2011 list of "top ten" gamma-ray sources. The top five sources in the Milky Way Galaxy: The Crab Nebula, W44, V407 Cygni, Pulsar PSR J0101-6422, 2FGL J0359.5+5410.

The top five sources outside the Milky Way Galaxy: Centaurus A, The Andromeda Galaxy (M31), The Cigar Galaxy (M82), Blazar PKS 0537-286, 2FGL J1305.0+1152.

A previous year 2009 "top ten" list of gamma-ray sources was created.

Ultraviolet Astronomy

Ultraviolet astronomy is the observation of electromagnetic radiation at ultraviolet wavelengths between approximately 10 and 320 nanometres; shorter wavelengths—higher energy photons—are studied by X-ray astronomy and gamma ray astronomy. Light at these wavelengths is absorbed by the Earth's atmosphere, so observations at these wavelengths must be performed from the upper atmosphere or from space.

Ultraviolet line spectrum measurements are used to discern the chemical composition, densities, and temperatures of the interstellar medium, and the temperature and composition of hot young stars. UV observations can also provide essential information about the evolution of galaxies.

A GALEX image of the spiral galaxy Messier 81 in ultraviolet light. Credit:GALEX/NASA/JPL-Caltech.

The ultraviolet Universe looks quite different from the familiar stars and galaxies seen in visible light. Most stars are actually relatively cool objects emitting much of their electromagnetic radiation in the visible or near-infrared part of the spectrum. Ultraviolet radiation is the signature of hotter objects, typically in the early and late stages of their evolution. If we could see the sky in ultraviolet light, most stars would fade in prominence. We would see some very young massive stars and some very old stars and galaxies, growing hotter and producing higher-energy radiation near their birth or death. Clouds of gas and dust would block our vision in many directions along the Milky Way.

The Hubble Space Telescope and FUSE have been the most recent major space telescopes to view the near and far UV spectrum of the sky, though other UV instruments have flown on sounding rockets and the Space Shuttle.

Charles Stuart Bowyer is generally given credit for starting this field.

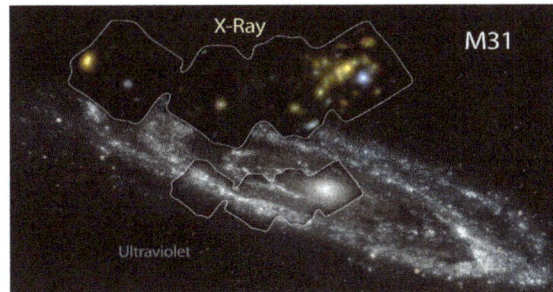

Andromeda Galaxy - in high-energy X-ray and ultraviolet light (released 5 January 2016).

Ultraviolet Space Telescopes

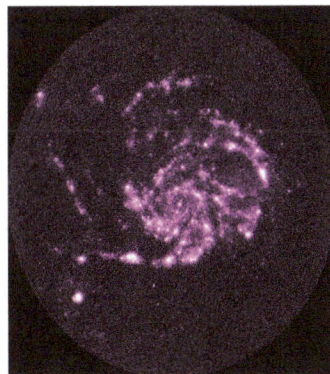

Astro 2 UIT captures M101 with ultraviolet shown in purple

- 🇺🇸- Far Ultraviolet Camera/Spectrograph on Apollo 16 (April 1972)
- 🇺🇸+ ESRO - TD-1A (135-286 nm; 1972–74)
- 🇺🇸- Orbiting Astronomical Observatory (#2:1968-73. #3:1972-81)
- 🟧- Orion 1 and Orion 2 Space Observatories (#1:1971; 200-380 nm spectra; #2:1973; 200-300 nm spectra)
- 🇺🇸+ 🇳🇱- Astronomical Netherlands Satellite (150-330 nm, 1974–76)
- 🇺🇸+ ESA - International Ultraviolet Explorer (115-320 nm spectra, 1978–96)
- 🟧- Astron-1 (1983–89; 150-350 nm)
- 🟧- Glazar 1 & 2 on Mir (in Kvant-1, 1987-2001)

- - EUVE (7-76 nm, 1992-2001)

- - FUSE (90.5-119.5 nm, 1999-2007)

- + ESA - Extreme ultraviolet Imaging Telescope (on SOHO imaging sun at 17.1, 19.5, 28.4, and 30.4 nm)

- - GALEX (135-280 nm, 2003-2013)

- + ESA - Hubble Space Telescope (Hubble STIS 1997–115–1030 nm) (Hubble WFC3 2009–200-1700 nm)

- - Swift Gamma-Ray Burst Mission (170–650 nm spectra, 2004--)

- - Hopkins Ultraviolet Telescope (flew in 1990 and 1995)

- - ROSAT XUV (17-210eV) (30-6 nm, 1990-1999)

- - Public Telescope (PST) (100-180 nm, Launch planned 2019)

- - Astrosat (130-530 nm, launched in September 2015)

Neutrino Astronomy

Neutrino astronomy is the branch of astronomy that observes astronomical objects with neutrino detectors in special observatories. Neutrinos are created as a result of certain types of radioactive decay, or nuclear reactions such as those that take place in the Sun, in nuclear reactors, or when cosmic rays hit atoms. Due to their weak interactions with matter, neutrinos offer a unique opportunity to observe processes that are inaccessible to optical telescopes.

The field of neutrino astronomy is still very much in its infancy – the only confirmed extraterrestrial sources so far are the Sun and supernova SN1987A.

History

Neutrinos were first recorded in 1956 by Clyde Cowan and Frederick Reines from a nuclear reactor. Their discovery was acknowledged with a Nobel Prize for physics in 1995.

In 1968, Raymond Davis, Jr. and John N. Bahcall successfully detected the first solar neutrinos in the Homestake experiment. Davis, along with Japanese physicist Masatoshi Koshiba were jointly awarded half of the 2002 Nobel Prize in Physics "for pioneering contributions to astrophysics, in particular for the detection of cosmic neutrinos (the other half went to Riccardo Giacconi for corresponding pioneering contributions which have led to the discovery of cosmic X-ray sources)."

This was followed by the first atmospheric neutrino detection in 1965 by two groups almost simultaneously. One was led by Frederick Reines who operated a liquid scintillator in the East Rand gold mine in South Africa at an 8.8 km water depth equivalent. The other was a Bombay-Osaka-Durham collaboration that operated in the Indian Kolar Gold Field mine at an equivalent water depth of 7.5 km. Although the KGF group detected neutrino candidates two months later than Reines, they were given formal priority due to publishing their findings two weeks earlier.

The first generation of undersea neutrino telescope projects began with the proposal by Moisey Markov in 1960 "...to install detectors deep in a lake or a sea and to determine the location of charged particles with the help of Cherenkov radiation."

The first underwater neutrino telescope began as the DUMAND project. DUMAND stands for Deep Underwater Muon and Neutrino Detector. The project began in 1976 and although it was eventually cancelled in 1995, it acted as a precursor to many of the following telescopes in the following decades.

The Baikal Neutrino Telescope is installed in the southern part of Lake Baikal in Russia. The detector is located at a depth of 1.1 km and began surveys in 1980. In 1993, it was the first to deploy three strings to reconstruct the muon trajectories as well as the first to record atmospheric neutrinos underwater.

AMANDA (Antarctic Muon And Neutrino Detector Array) used the 3 km thick ice layer at the South Pole and was located several hundred meters from the Amundsen-Scott station. Holes 60 cm in diameter were drilled with pressurized hot water in which strings with optical modules were deployed before the water refroze. The depth proved to be insufficient to be able to reconstruct the trajectory due to the scattering of light on air bubbles. A second group of 4 strings were added in 1995/96 to a depth of about 2000 m that was sufficient for track reconstruction. The AMANDA array was subsequently upgraded until January 2000 when it consisted of 19 strings with a total of 667 optical modules at a depth range between 1500 m and 2000 m. AMANDA would eventually be the predecessor to IceCube in 2005.

21st Century

After the decline of DUMAND the participating groups split into three branches to explore deep sea options in the Mediterranean Sea. ANATES was anchored to the sea floor in the region off Toulon at the French Mediterranean coast. It consists of 12 strings, each carrying 25 "storeys" equipped with three optical modules, an electronic container, and calibration devices down to a maximum depth of 2475 m.

NEMO (NEutrino Mediterranean Observatory) was pursued by Italian groups to investigate the feasibility of a cubic-kilometer scale deep-sea detector. A suitable site at a depth of 3.5 km about 100 km off Capo Passero at the South-Eastern coast of Sicily has been identified. From 2007-2011 the first prototyping phase tested a "mini-tower" with 4 bars deployed for several weeks near Catania at a depth of 2 km. The second phase as well as plans to deploy the full-size prototype tower will be pursued in the KM3NeT framework.

The NESTOR Project was installed in 2004 to a depth of 4 km and operated for one month until a failure of the cable to shore forced it to be terminated. The data taken still successfully demonstrated the detector's functionality and provided a measurement of the atmospheric muon flux. The proof of concept will be implemented in the KM3Net framework.

The second generation of deep-sea neutrino telescope projects reach or even exceed the size originally conceived by the DUMAND pioneers. IceCube, located at the South Pole and incorporating its predecessor AMANDA, was completed in December 2010. It currently consists of 5160 digital optical modules installed on 86 strings at depths of 1450 to 2550 m in the Antarctic ice. The KM3NeT in the Mediterranean Sea and the GVD are in their preparatory/prototyping phase. IceCube instruments 1 km³ of ice. GVD is also planned to cover 1 km³ but at a much higher energy threshold. KM3NeT is planned to cover several km³. Both KM3NeT and GVD could be completed by 2017 and it is expected that all three will form a global neutrino observatory.

Detection Methods

Since neutrinos interact only very rarely with matter, the enormous flux of solar neutrinos racing through the Earth is sufficient to produce only 1 interaction for 10^{36} target atoms, and each interaction produces only a few photons or one transmuted atom. The observation of neutrino interactions requires a large detector mass, along with a sensitive amplification system.

Given the very weak signal, sources of background noise must be reduced as much as possible. The detectors must be shielded by a large shield mass, and so are constructed deep underground, or underwater. They record upward going muons in charged current muon neutrino interactions. Upward because no other known particle can traverse the entire Earth. The detector must be at least 1 km deep to suppress downward traveling muons, and are subject to an irreducible background of extraterrestrial neutrinos interacting in the Earth's atmosphere. This background also provides a standard calibration source. Sources of radioactive isotopes must also be controlled as they produce energetic particles when they decay. The detectors consist of an array of photomultiplier tubes (PMTs) housed in transparent pressure spheres which are suspended in a large volume of water or ice. The PMTs record the arrival time and amplitude of the Cherenkov light emitted by muons or particle cascades. The trajectory can then usually be reconstructed by triangulation if at least three "strings" are used to detect the events.

Applications

When astronomical bodies, such as the Sun, are studied using light, only the surface of the object can be directly observed. Any light produced in the core of a star will interact with gas particles in the outer layers of the star, taking hundreds of thousands of years to make it to the surface, making it impossible to observe the core directly. Since neutrinos are also created in the cores of stars (as a result of stellar fusion), the core can be observed using neutrino astronomy. Other sources of neutrinos- such as neutrinos released by supernovae- have been detected. There are currently goals to detect neutrinos from other sources, such as Active Galactic Nuclei (AGN), as well as Gamma-ray bursts and Starburst galaxies. Neutrino astronomy may also indirectly detect dark matter.

Visible-light Astronomy

Visible-light astronomy encompasses a wide variety of observations via telescopes that are sensitive in the range of visible light (optical telescopes). It includes *imaging*, where a picture of some sort is made of the object; *photometry*, where the amount of light coming from an object is measured, *spectroscopy*, where the distribution of that light with respect to its wavelength is measured, and *polarimetry* where the polarisation state of that light is measured. An example of spectroscopy is the study of spectral lines to understand of what kind of matter light is going through. Visible astronomy also includes looking up at night (skygazing). Visible-light astronomy is part of optical astronomy, and differs from astronomies based on invisible types of light in the electromagnetic radiation spectrum, such as radio waves, infrared waves, ultraviolet waves, X-ray waves and gamma-ray waves.

Beginning

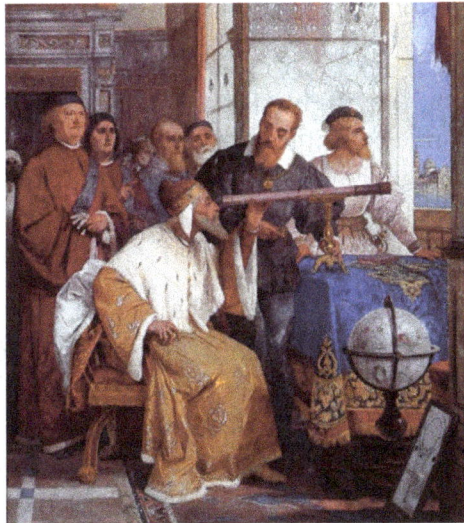

Fresco by Giuseppe Bertini depicting Galileo showing the Doge of Venice how to use the telescope

Based only on uncertain descriptions of the first practical telescope, invented by Hans Lippershey in the Netherlands in 1608, Galileo, in the following year, made a telescope with about 3x magnification. He later made improved versions with up to about 30x magnification. With a Galilean telescope the observer could see magnified, upright images on the earth—it was what is commonly known as a terrestrial telescope or a spyglass. He could also use it to observe the sky; for a time he was one of those who could construct telescopes good enough for that purpose. On 25 August 1609, he demonstrated one of his early telescopes, with a magnification of about 8 or 9, to Venetian lawmakers. His telescopes were also a profitable sideline for Galileo selling them to merchants who found them useful both at sea and as items of trade. He published his initial telescopic astronomical observations in March 1610 in a brief treatise entitled *Sidereus Nuncius* (*Starry Messenger*).

Effect of Ambient Brightness

The visibility of celestial objects in the night sky is affected by light pollution. The presence of the Moon in the night sky has historically hindered astronomical observation by increasing the amount of ambient lighting. With the advent of artificial light sources, however, light pollution has been a growing problem for viewing the night sky. Special filters and modifications to light fixtures can help to alleviate this problem, but for the best views, both professional and amateur optical astronomers seek viewing sites located far from major urban areas.

Nuclear Astrophysics

Nuclear astrophysics is an interdisciplinary branch of physics involving close collaboration among researchers in various subfields of nuclear physics and astrophysics, with significant emphasis in areas such as stellar modeling, measurement and theoretical estimation of nuclear reaction rates, cosmology, cosmochemistry, gamma ray, optical and X-ray astronomy, and extending our knowledge about nuclear lifetimes and masses. In general terms, nuclear astrophysics aims to understand the origin of the chemical elements and the energy generation in stars.

History

The basic principles of explaining the origin of the elements and the energy generation in stars were laid down in the theory of nucleosynthesis which came together in the late 1950s from the seminal works of Burbidge, Burbidge, Fowler, and Hoyle in a famous paper and independently by Cameron. Fowler is largely credited with initiating the collaboration between astronomers, astrophysicists, and experimental nuclear physicists which is what we now know as nuclear astrophysics, winning the Nobel Prize for this in 1983.

The basic tenets of nuclear astrophysics are that only isotopes of hydrogen and helium (and traces of lithium, beryllium, and boron) can be formed in a homogeneous big bang model and all other elements are formed in stars. The conversion of nuclear mass to radiative energy (by merit of Einstein's famous mass-energy relation in relativity) is the source of energy which allows stars to shine for up to billions of years. Many notable physicists of the 19th century, such as Mayer, Waterson, von Helmholtz, and Lord Kelvin, postulated that the Sun radiates thermal ener-gy based on converting gravitational potential energy into heat. The lifetime of the Sun under such a model can be calculated relatively easily using the virial theorem, yielding around 19 million years, an age that was not consistent with the interpretation of geological records or the then recently proposed theory of biological evolution. A back-of-the-envelope calculation indicates that if the Sun consisted entirely of a fossil fuel like coal, a source of energy familiar to many people, considering the rate of thermal energy emission, then the Sun would have a lifetime of merely four or five thousand years, which is not even consistent with records of human civilization. The now discredited hypothesis that gravitational contraction is the Sun's primary source of energy was, however, reasonable before the advent of modern physics; radioactivity itself was not discovered by Becquerel until 1895 Besides the prerequisite knowledge of the atomic nucleus, a proper understanding of stellar energy is not possible without the theories of relativity and quantum mechanics.

After Aston demonstrated that the mass of helium is less than four times the mass of the proton, Eddington proposed that in the core of the Sun, through an unknown process, hydrogen was transmuted into helium, liberating energy. 20 years later, Bethe and von Weizsäcker independently derived the CN cycle, the first known nuclear reaction cycle which can accomplish this transmutation; however, it is now understood that the Sun's primary energy source is the pp-chains, which can occur at much lower energies and are much slower than catalytic hydrogen fusion. The time-lapse between Eddington's proposal and the derivation of the CN cycle can mainly be attributed to an incomplete understanding of nuclear structure, and a proper understanding of nucleosynthetic processes was not possible until Chadwick discovered the neutron in 1932 and a contemporary theory of beta decay developed. Nuclear physics gives a self-consistent picture of the energy source for the Sun and its subsequent lifetime, as the age of the Solar System derived from meteoritic abundances of lead and uranium isotopes is about 4.5 billion years. A star the mass of the Sun has enough nuclear fuel to allow for core hydrogen burning on the main sequence of the HR-diagram via the pp-chains for about 9 billion years, a lifetime primarily set by the extremely slow production of deuterium,

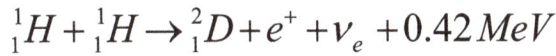

$$ {}^1_1H + {}^1_1H \rightarrow {}^2_1D + e^+ + \nu_e + 0.42\,MeV $$

which is governed by the nuclear weak force.

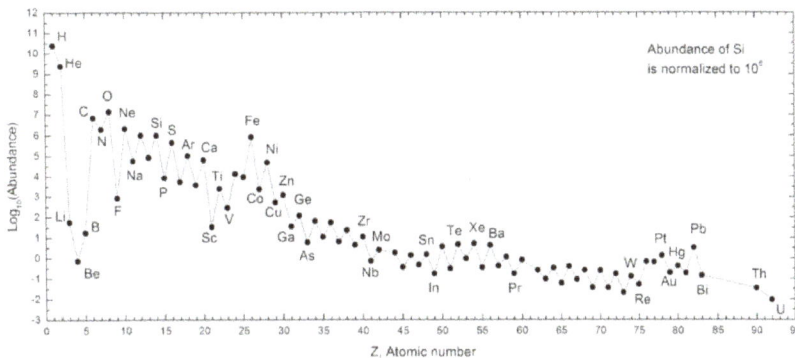

Abundances of the chemical elements in the Solar System. Hydrogen and helium are most common, residuals within the paradigm of the Big Bang. The next three elements (Li, Be, B) are rare because they are poorly synthesized in the Big Bang and also in stars. The two general trends in the remaining stellar-produced elements are: (1) an alternation of abundance of elements according to whether they have even or odd atomic numbers, and (2) a general decrease in abundance, as elements become heavier. Within this trend is a peak at abundances of iron and nickel, which is especially visible on a logarithmic graph spanning fewer powers of ten, say between logA=2 (A=100) and logA=6 (A=1,000,000).

Predictions

The theory of stellar nucleosynthesis reproduces the chemical abundances observed in the Solar System and galaxy, which from hydrogen to uranium, show an extremely varied distribution spanning twelve orders of magnitude (one trillion). While impressive, these data were used to formulate the theory, and a scientific theory must be predictive in order to have any merit. The theory of stellar nucleosynthesis has been well-tested by observation and experiment since the theory was first formulated.

The theory predicted the observation of technetium (the lightest chemical element with no stable isotopes) in stars, observation of galactic gamma-emitters such as ^{26}Al and ^{44}Ti, observation of solar neutrinos, and observation of neutrinos from supernova 1987a. These observations have far-reaching implications. ^{26}Al has a lifetime a bit less than one million years, which is very short on a galactic timescale, proving that nucleosynthesis is an ongoing process even in our own time. Work which led to the discovery of neutrino oscillation, implying a non-zero mass for the neutrino and thus not predicted by the Standard Model of particle physics, was motivated by a solar neutrino flux about three times lower than expected, which was a long-standing concern in the nuclear astrophysics community such that it was colloquially known simply as the Solar neutrino problem. The observable neutrino flux from nuclear reactors is much larger than that of the Sun, and thus Davis and others were primarily motivated to look for solar neutrinos for astronomical reasons.

Future Work

Although the foundations of the science are bona fide, there are still many remaining open questions. A few of the long-standing issues are helium fusion (specifically the $^{12}C(\alpha,\gamma)^{16}O$ reaction), the astrophysical site of the r-process, anomalous lithium abundances in Population III stars, and the explosion mechanism in core-collapse supernovae.

References

- Orchiston, W. (2005). "Dr Elizabeth Alexander: First Female Radio Astronomer". The New Astronomy: Opening the Electromagnetic Window and Expanding Our View of Planet Earth. Astrophysics and Space Science Library. 334. pp. 71–92. doi:10.1007/1-4020-3724-4_5. ISBN 978-1-4020-3723-8.

- Russell CT (2001). "Solar wind and interplanetary magnetic filed: A tutorial". In Song, Paul; Singer, Howard J.; Siscoe, George L. Space Weather (Geophysical Monograph) (PDF). American Geophysical Union. pp. 73–88. ISBN 978-0-87590-984-4.

- Carlino, G.; D'Ambrosio, G.; Merola, L.; Paolucci, P.; Ricciardi, G. (16 September 2008). IFAE 2007: Incontri di Fisica delle Alte Energie Italian Meeting on High Energy Physics. Springer Science & Business Media. p. 245. ISBN 978-88-470-0747-5. Retrieved 21 August 2014.

- Paredes, Josep M.; Reimer, Olaf; Torres, Diego F. (17 July 2007). The Multi-Messenger Approach to High-Energy Gamma-Ray Sources: Third Workshop on the Nature of Unidentified High-Energy Sources. Springer. p. 180. ISBN 978-1-4020-6118-9. Retrieved 21 August 2014.

- Krieg, Uwe (2008). Siegfried Röser, eds. Reviews in Modern Astronomy, Cosmic Matter. 20. WILEY-VCH. p. 191. ISBN 978-3-527-40820-7. Retrieved 2010-11-14.

- Barnes, C. A.; Clayton, D. D.; Schramm, D. N., eds. (1982), Essays in Nuclear Astrophysics, Cambridge University Press, ISBN 0-52128-876-2

- Shields, Gregory A. (1999). "A brief history of AGN". The Publications of the Astronomical Society of the Pacific. 111 (760): 661–678. arXiv:astro-ph/9903401. Bibcode:1999PASP..111..661S. doi:10.1086/316378. Retrieved 3 October 2014.

- Gelmini, G. B.; Kusenko, A.; Weiler, T. J. (18 May 2010). "Through Neutrino Eyes: Ghostly Particles Become Astronomical Tools". Scientific American. Retrieved 2013-11-28.

- for example, supernova SN 1987A emitted an "afterglow" of gamma-ray photons from the decay of newly made radioactive cobalt-56 ejected into space in a cloud, by the explosion. "The Electromagnetic Spectrum - Gamma-rays". NASA. Retrieved 2010-11-14.

- Morrison, Philip (March 16, 1958). "On gamma-ray astronomy". Il Nuovo Cimento (1955-1965). 7 (6): 858–865. doi:10.1007/BF02745590. Retrieved 2010-11-14.

- "Cosmic Rays Hunted Down: Physicists Closing in on Origin of Mysterious Particles". ScienceDaily. Dec 7, 2009. Retrieved 2010-11-14.

3

Essential Concepts of Astrophysics

Astrophysics is best understood in confluence with the major topics listed in the following chapter. The major concepts of astrophysics such as nebula, active galactic nucleus, Hayashi track, and the Eddington number are explained. This chapter serves as a source to understand the major concepts related to astrophysics.

Astronomical Constant

An astronomical constant is a physical constant used in astronomy. Formal sets of constants, along with recommended values, have been defined by the International Astronomical Union (IAU) several times: in 1964 and in 1976 (with an update in 1994). In 2009 the IAU adopted a new current set, and recognizing that new observations and techniques continuously provide better values for these constants, they decided to not fix these values, but have the Working Group on Numerical Standards continuously maintain a set of Current Best Estimates . The set of constants is widely reproduced in publications such as the *Astronomical Almanac* of the United States Naval Observatory and HM Nautical Almanac Office.

Besides the IAU list of units and constants, also the International Earth Rotation and Reference Systems Service defines constants relevant to the orientation and rotation of the Earth, in its technical notes .

The IAU system of constants defines a system of astronomical units for length, mass and time (in fact, several such systems), and also includes constants such as the speed of light and the constant of gravitation which allow transformations between astronomical units and SI units. Slightly different values for the constants are obtained depending on the frame of reference used. Values quoted in barycentric dynamical time (TDB) or equivalent time scales such as the *Teph* of the Jet Propulsion Laboratory ephemerides represent the mean values that would be measured by an observer on the Earth's surface (strictly, on the surface of the geoid) over a long period of time. The IAU also recommends values in SI units, which are the values which would be measured (in proper length and proper time) by an observer at the barycentre of the Solar System: these are obtained by the following transformations:

$$\tau_A(\text{SI}) = (1+L_B)^{\frac{1}{3}}\tau_A(\text{TDB})$$

$$GE(\text{SI}) = (1+L_B)GE(\text{TDB})$$

$$GS(\text{SI}) = (1+L_B)GS(\text{TDB})$$

Astronomical System of Units

The astronomical unit of time is a time interval of one day (D) of 86400 seconds. The astronomical unit of mass is the mass of the Sun (S). The astronomical unit of length is that length (A) for which the Gaussian gravitational constant (k) takes the value 0.017202 098 95 when the units of measurement are the astronomical units of length, mass and time.

Table of Astronomical Constants

Quantity	Symbol	Value	Relative uncertainty
Defining constants			
Gaussian gravitational constant	k	0.017 202 098 95 $A^{3/2}\,S^{-1/2}\,D^{-1}$	defined
Speed of light	c	299 792 458 m s^{-1}	defined
Mean ratio of the TT second to the TCG second	$1 - L_G$	$1 - 6.969\,290\,134\times10^{-10}$	defined
Mean ratio of the TCB second to the TDB second	$1 - L_B$	$1 - 1.550\,519\,767\,72\times10^{-8}$	defined
Primary constants			
Mean ratio of the TCB second to the TCG second	$1 - L_C$	$1 - 1.480\,826\,867\,41\times10^{-8}$	1.4×10^{-9}
Light-time for unit distance	τ_A	499.004 786 3852 s	4.0×10^{-11}
Equatorial radius for Earth	a_e	$6.378\,1366\times10^{6}$ m	1.6×10^{-8}
Potential of the geoid	W_0	$6.263\,685\,60\times10^{7}$ m^2 s^{-2}	8.0×10^{-9}
Dynamical form-factor for Earth	J_2	0.001 082 6359	9.2×10^{-8}
Flattening factor for Earth	$1/f$	0.003 352 8197 = 1/298.256 42	3.4×10^{-8}
Geocentric gravitational constant	GE	$3.986\,004\,391\times10^{14}$ m^3 s^{-2}	2.0×10^{-9}
Constant of gravitation	G	$6.673\,84\times10^{-11}$ m^3 kg^{-1} s^{-2}	1.2×10^{-4}
Ratio of mass of Moon to mass of Earth	μ	0.012 300 0383 = 1/81.300 56	4.0×10^{-8}
General precession in longitude, per Julian century, at standard epoch 2000	ρ	5029.796 195″	*
Obliquity of the ecliptic, at standard epoch 2000	ε	23° 26′ 21.406″	*
Derived constants			
Constant of nutation, at standard epoch 2000	N	9.205 2331″	*
Unit distance = $c\tau_A$	A	149 597 870 691 m	4.0×10^{-11}
Solar parallax = $\arcsin(a_e/A)$	$\pi_{\odot}$	8.794 1433″	1.6×10^{-8}
Constant of aberration, at standard epoch 2000	κ	20.495 52″	
Heliocentric gravitational constant = $A^3 k^2/D^2$	GS	$1.327\,2440\times10^{20}$ m^3 s^{-2}	3.8×10^{-10}
Ratio of mass of Sun to mass of Earth = $(GS)/(GE)$	S/E	332 946.050 895	
Ratio of mass of Sun to mass of (Earth + Moon)	$\dfrac{(S/E)}{(1+\mu)}$	328 900.561 400	
Mass of Sun = $(GS)/G$	S	1.9818×10^{30} kg	1.0×10^{-4}
System of planetary masses: Ratios of mass of Sun to mass of planet[10]			
Mercury		6 023 600	

Venus		408 523.71	
Earth + Moon		328 900.561 400	
Mars		3 098 708	
Jupiter		1047.3486	
Saturn		3497.898	
Uranus		22 902.98	
Neptune		19 412.24	
Pluto		135 200 000	
Other constants (outside the formal IAU System)			
Parsec = A/tan(1")	pc	$3.085\,677\,581\,28{\times}10^{16}$ m	$4.0{\times}10^{-11}$
Light-year = 365.25cD	ly	$9.460\,730\,472\,5808{\times}10^{15}$ m	defined
Hubble constant	H_0	$70.1\,\mathrm{km\,s^{-1}\,Mpc^{-1}}$	0.019
Solar luminosity	L_0	$3.939{\times}10^{26}$ W $= 2.107{\times}10^{-15}\,\mathrm{S\,D^{-1}}$	variable, ±0.1%

Notes

* The theories of precession and nutation have advanced since 1976, and these also affect the definition of the ecliptic. The values here are appropriate for the older theories, but additional constants are required for current models.

† The definitions of these derived constants have been taken from the references cited, but the values have been recalculated to take account of the more precise values of the primary constants cited in the table.

Nebula

The "Pillars of Creation" from the Eagle Nebula. Evidence from the Spitzer Telescope suggests that the pillars may already have been destroyed by a supernova explosion, but the light showing us the destruction will not reach the Earth for another millennium.

A nebula (Latin for "cloud"; pl. nebulae, nebulæ, or nebulas) is an interstellar cloud of dust, hydrogen, helium and other ionized gases. Originally, *nebula* was a name for any diffuse astronomical object, including galaxies beyond the Milky Way. The Andromeda Galaxy, for instance, was once referred to as the *Andromeda Nebula* (and spiral galaxies in general as "spiral nebulae") before the true nature of galaxies was confirmed in the early 20th century by Vesto Slipher, Edwin Hubble and others.

Most nebulae are of vast size, even millions of light years in diameter. Contrary to fictional depictions where starships hide in nebulae as thick as cloud banks, in reality a nebula that is barely visible to the human eye from Earth would appear larger, but no brighter, from close by. The Orion Nebula, the brightest nebula in the sky that occupies a region twice the diameter of the full Moon, can be viewed with the naked eye but was missed by early astronomers. Although denser than the space surrounding them, most nebulae are far less dense than any vacuum created on Earth – a nebular cloud the size of the Earth would have a total mass of only a few kilograms. Many nebulae are visible due to their fluorescence caused by the embedded hot stars, while others are so diffuse they can only be detected with long exposures and special filters. Some nebulae are variably illuminated by T Tauri variable stars. Nebulae are often star-forming regions, such as in the "Pillars of Creation" in the Eagle Nebula. In these regions the formations of gas, dust, and other materials "clump" together to form denser regions, which attract further matter, and eventually will become dense enough to form stars. The remaining material is then believed to form planets and other planetary system objects.

Observational History

Portion of the Carina Nebula

Around 150 AD, Claudius Ptolemaeus (Ptolemy) recorded, in books VII-VIII of his *Almagest*, five stars that appeared nebulous. He also noted a region of nebulosity between the constellations Ursa Major and Leo that was not associated with any star. The first true nebula, as distinct from a star cluster, was mentioned by the Persian astronomer, Abd al-Rahman al-Sufi, in his *Book of Fixed Stars* (964). He noted "a little cloud" where the Andromeda Galaxy is located. He also cataloged the Omicron Velorum star cluster as a "nebulous star" and other nebulous objects, such as Brocchi's Cluster. The supernova that created the Crab Nebula, the SN 1054, was observed by Arabic and Chinese astronomers in 1054.

In 1610, Nicolas-Claude Fabri de Peiresc discovered the Orion Nebula using a telescope. This nebula was also observed by Johann Baptist Cysat in 1618. However, the first detailed study of the Orion Nebula was not performed until 1659 by Christiaan Huygens, who also believed himself to be the first person to discover this nebulosity.

In 1715, Edmund Halley published a list of six nebulae. This number steadily increased during the century, with Jean-Philippe de Cheseaux compiling a list of 20 (including eight not previously known) in 1746. From 1751–53, Nicolas Louis de Lacaille cataloged 42 nebulae from the Cape of Good Hope, most of which were previously unknown. Charles Messier then compiled a catalog of 103 "nebulae" (now called Messier objects, which included what are now known to be galaxies) by 1781; his interest was detecting comets, and these were objects that might be mistaken for them.

The number of nebulae was then greatly expanded by the efforts of William Herschel and his sister Caroline Herschel. Their *Catalogue of One Thousand New Nebulae and Clusters of Stars* was published in 1786. A second catalog of a thousand was published in 1789 and the third and final catalog of 510 appeared in 1802. During much of their work, William Herschel believed that these nebulae were merely unresolved clusters of stars. In 1790, however, he discovered a star surrounded by nebulosity and concluded that this was a true nebulosity, rather than a more distant cluster.

Beginning in 1864, William Huggins examined the spectra of about 70 nebulae. He found that roughly a third of them had the emission spectrum of a gas. The rest showed a continuous spectrum and thus were thought to consist of a mass of stars. A third category was added in 1912 when Vesto Slipher showed that the spectrum of the nebula that surrounded the star Merope matched the spectra of the Pleiades open cluster. Thus the nebula radiates by reflected star light.

About 1922, following the Great Debate, it had become clear that many "nebulae" were in fact galaxies far from our own.

Slipher and Edwin Hubble continued to collect the spectra from many diffuse nebulae, finding 29 that showed emission spectra and 33 that had the continuous spectra of star light. In 1922, Hubble announced that nearly all nebulae are associated with stars, and their illumination comes from star light. He also discovered that the emission spectrum nebulae are nearly always associated with stars having spectral classifications of B1 or hotter (including all O-type main sequence stars), while nebulae with continuous spectra appear with cooler stars. Both Hubble and Henry Norris Russell concluded that the nebulae surrounding the hotter stars are transformed in some manner.

Formation

Many nebulae or stars form from the gravitational collapse of gas in the interstellar medium. As the material collapses under its own weight, massive stars may form in the center, and their ultraviolet radiation ionizes the surrounding gas, making it visible at optical wavelengths. Examples of these types of nebulae are the Rosette Nebula and the Pelican Nebula. The size of these nebulae, known as H II regions, varies depending on the size of the original cloud of gas. New stars are formed in the nebulae. The formed stars are sometimes known as a young, loose cluster.

The Triangulum Emission Garren Nebula NGC 604

Other nebulae form as the result of supernova explosions; the death throes of massive, short-lived stars. The materials thrown off from the supernova explosion are then ionized by the energy and the compact object that its core produces. One of the best examples of this is the Crab Nebula, in Taurus. The supernova event was recorded in the year 1054 and is labelled SN 1054. The compact object that was created after the explosion lies in the center of the Crab Nebula and its core is now a neutron star.

Still other nebulae form as planetary nebulae. This is the final stage of a low-mass star's life, like Earth's Sun. Stars with a mass up to 8–10 solar masses evolve into red giants and slowly lose their outer layers during pulsations in their atmospheres. When a star has lost enough material, its temperature increases and the ultraviolet radiation it emits can ionize the surrounding nebula that it has thrown off. Our Sun will produce a planetary nebula and its core will remain behind in the form of white dwarf.

Types of Nebulae

| Herbig–Haro object HH 161 and HH 164. | The Omega Nebula, an example of an emission nebula | The Horsehead Nebula, an example of a dark nebula. |

The Cat's Eye Nebula, an example of a planetary nebula.	The Red Rectangle Nebula, an example of a protoplanetary nebula.	The delicate shell of SNR B0509-67.5

Classical Types

Objects named nebulae belong to four major groups. Before their nature was understood, galaxies ("spiral nebulae") and star clusters too distant to be resolved as stars were also classified as nebulae, but no longer are.

- H II regions, large diffuse nebulae containing ionized hydrogen
- Planetary nebulae
- Supernova remnant (e.g., Crab Nebula)
- Dark nebula
- Not all cloud-like structures are named nebulae; Herbig–Haro objects are an example.

Diffuse Nebulae

The Carina Nebula is a diffuse nebula

Most nebulae can be described as diffuse nebulae, which means that they are extended and contain no well-defined boundaries. Diffuse nebulae can be divided into emission nebula, reflection nebulae and "dark nebulae." In visible light nebulae may be divided into emission nebulae that emit spectral line radiation from excited or ionized gas (mostly ionized hydrogen); they are often called HII regions (the term "HII" refers to ionized hydrogen). Reflection nebulae are visible primarily due to the light they reflect. Reflection nebulae themselves do not emit significant amounts of visible light, but are near stars and reflect light from them. Similar nebulae not illuminated by stars do not exhibit visible radiation, but may be detected as opaque clouds blocking light from luminous objects behind them; they are called "dark nebulae".

Although these nebulae have different visibility at optical wavelengths, they are all bright sources of infrared emission, chiefly from dust within the nebulae.

Planetary Nebulae

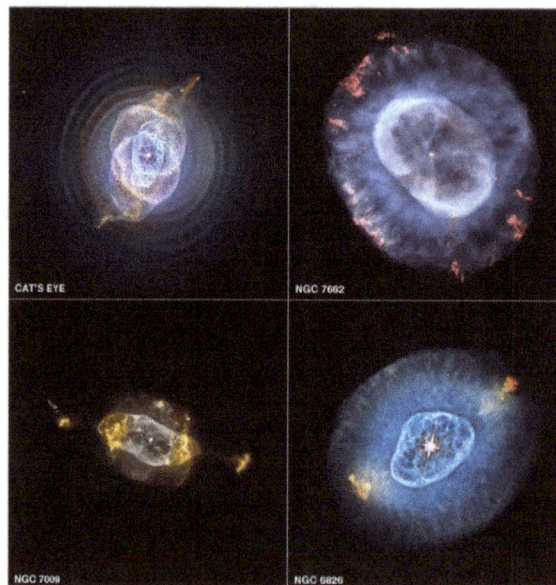

Four different planetary nebulae

Planetary nebulae form when low-mass asymptotic giant branch stars nova. A star that novas pushes the outer layers of the star's mass outward forming gaseous shells, while leaving behind the star's core in the form of a white dwarf. The hot white dwarf illuminates the expelled gases producing emission nebulae with spectra similar to those of emission nebulae found in star formation regions. Technically they are HII regions, because most hydrogen will be ionized, but they are denser and more compact than the nebulae found in star formation regions. Planetary nebulae were given their name by the first astronomical observers who were initially unable to distinguish them from planets, and who tended to confuse them with planets, which were of more interest to them. Our Sun is expected to spawn a planetary nebula about 12 billion years after its formation.

Protoplanetary Nebula

A protoplanetary nebula (PPN) is an astronomical object which is at the short-lived episode during a star's rapid stellar evolution between the late asymptotic giant branch (LAGB) phase and the following planetary nebula (PN) phase. During the AGB phase, the star undergoes mass loss, emitting a circumstellar shell of hydrogen gas. When this phase comes to an end, the star enters the PPN phase.

The PPN is energized by the central star, causing it to emit strong infrared radiation and become a reflection nebula. Collimated stellar winds from the central star shape and shock the shell into an axially symmetric form, while producing a fast moving molecular wind. The exact point when a PPN becomes a planetary nebula (PN) is defined by the temperature of the central star. The PPN phase continues until the central star reaches a temperature of 30,000 K, after which it is hot enough to ionize the surrounding gas.

Supernova Remnants

The Crab Nebula, an example of a supernova remnant

A supernova occurs when a high-mass star reaches the end of its life. When nuclear fusion in the core of the star stops, the star collapses. The gas falling inward either rebounds or gets so strongly heated that it expands outwards from the core, thus causing the star to explode. The expanding shell of gas forms a supernova remnant, a special diffuse nebula. Although much of the optical and X-ray emission from supernova remnants originates from ionized gas, a great amount of the radio emission is a form of non-thermal emission called synchrotron emission. This emission originates from high-velocity electrons oscillating within magnetic fields.

Notable Named Nebulae

- Ant Nebula
- Barnard's Loop
- Boomerang Nebula
- Cat's Eye Nebula
- Crab Nebula
- Eagle Nebula
- Eskimo Nebula

- Eta Carinae Nebula
- Fox Fur Nebula
- Helix Nebula
- Hourglass Nebula
- Horsehead Nebula
- Lagoon Nebula
- Orion Nebula
- Pelican Nebula
- Red Square Nebula
- Ring Nebula
- Rosette Nebula
- Tarantula Nebula

Nebula Catalogs

- Gum catalog
- RCW Catalogue
- Sharpless catalog
- Caldwell Catalogue

Accretion (Astrophysics)

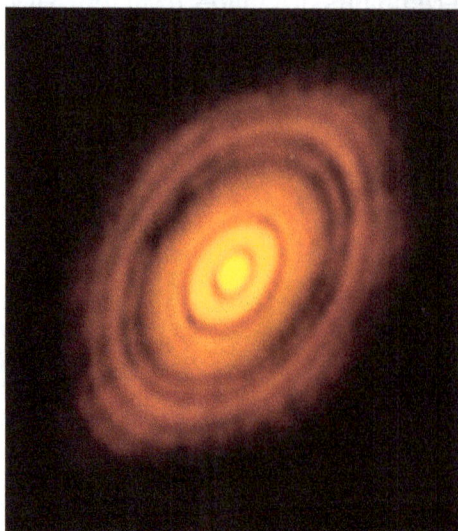

Atacama Large Millimeter Array image of HL Tauri, a protoplanetary disk

In astrophysics, accretion is the accumulation of particles into a massive object by gravitationally attracting more matter, typically gaseous matter, in an accretion disk. Most astronomical objects, such as galaxies, stars, and planets, are formed by accretion processes.

Overview

The idea proposed in the 19th century that Earth and the other terrestrial planets formed from meteoric material was developed in a quantitative way in 1969 by Viktor Safronov. He calculated, in detail, the different stages of terrestrial planet formation. Since then, the theory has been further developed using intensive numerical simulations to study planetesimal accumulation.

Stars form by the gravitational collapse of interstellar gas. Prior to collapse, this gas is mostly in the form of molecular clouds, such as the Orion Nebula. As the cloud collapses, losing potential energy, it heats up, gaining kinetic energy, and the conservation of angular momentum ensures that the cloud forms a flatted disk—the accretion disk.

Accretion of Galaxies

Further information: Protogalaxy

A few hundred thousand years after the Big Bang, the Universe cooled to the point where atoms could form. As the Universe continued to expand and cool, the atoms lost enough kinetic energy, and dark matter coalesced sufficiently, to form protogalaxies. As further accretion occurred, galaxies formed. Indirect evidence is widespread. Galaxies grow through mergers and smooth gas accretion. Accretion also occurs inside galaxies, forming stars.

Accretion of Stars

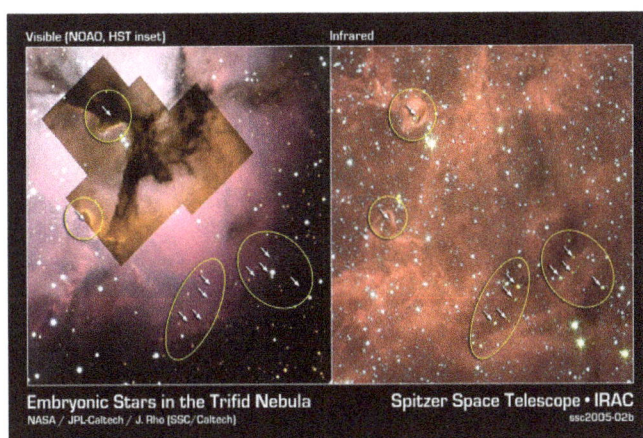

Visible (NOAO, HST inset)   Infrared

Embryonic Stars in the Trifid Nebula   Spitzer Space Telescope • IRAC
NASA / JPL-Caltech / J. Rho (SSC/Caltech)   ssc2005-02b

The visible-light (left) and infrared (right) views of the Trifid Nebula, a giant star-forming cloud of gas and dust located 5,400 light-years (1,700 pc) away in the constellation Sagittarius

Stars are thought to form inside giant clouds of cold molecular hydrogen—giant molecular clouds of roughly 300,000 $M_\odot$ and 65 light-years (20 pc) in diameter. Over millions of years, giant molecular clouds are prone to collapse and fragmentation. These fragments then form small, dense cores, which in turn collapse into stars. The cores range in mass from a fraction to several times that of the Sun and are called protostellar (protosolar) nebulae. They possess diameters of 2,000–20,000 astronomical units (0.01–0.1 pc) and a particle number density of roughly 10,000 to 100,000/cm³ (160,000 to 1,600,000/cu in). Compare it with the particle number density of the air at the sea level—2.8×10^{19}/cm³ (4.6×10^{20}/cu in).

The initial collapse of a solar-mass protostellar nebula takes around 100,000 years. Every nebula begins with a certain amount of angular momentum. Gas in the central part of the nebula, with relatively low angular momentum, undergoes fast compression and forms a hot hydrostatic (non-contracting) core containing a small fraction of the mass of the original nebula. This core forms the seed of what will become a star. As the collapse continues, conservation of angular momentum dictates that the rotation of the infalling envelope accelerates, which eventually forms a disk.

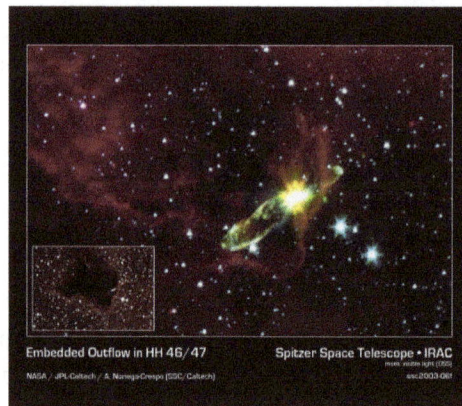

Infrared image of the molecular outflow from an otherwise hidden newborn star HH 46/47

As the infall of material from the disk continues, the envelope eventually becomes thin and transparent and the young stellar object (YSO) becomes observable, initially in far-infrared light and later in the visible. Around this time the protostar begins to fuse deuterium. If the protostar is sufficiently massive (above 80 M_J), hydrogen fusion follows. Otherwise, if its mass is too low, the object becomes a brown dwarf. This birth of a new star occurs approximately 100,000 years after the collapse begins. Objects at this stage are known as Class I protostars, which are also called young T Tauri stars, evolved protostars, or young stellar objects. By this time, the forming star has already accreted much of its mass; the total mass of the disk and remaining envelope does not exceed 10–20% of the mass of the central YSO.

When the lower-mass star in a binary system enters an expansion phase, its outer atmosphere may fall onto the compact star, forming an accretion disk

At the next stage, the envelope completely disappears, having been gathered up by the disk, and the protostar becomes a classical T Tauri star. The latter have accretion disks and continue to accrete hot gas, which manifests itself by strong emission lines in their spectrum. The former do not possess accretion disks. Classical T Tauri stars evolve into weakly lined T Tauri stars. This happens after about 1 million years. The mass of the disk around a classical T Tauri star is about 1–3% of the stellar mass, and it is accreted at a rate of 10^{-7} to 10^{-9} $M_\odot$ per year. A pair of bipolar jets is usually present as well. The accretion explains all peculiar properties of classical T Tauri stars: strong flux in the emission lines (up to 100% of the intrinsic luminosity of the star), magnetic activity, photometric variability and jets. The emission lines actually form as the accreted gas hits the "surface" of the star, which happens around its magnetic poles. The jets are byproducts of accretion: they carry away excessive angular momentum. The classical T Tauri stage lasts about 10 million years. The disk eventually disappears due to accretion onto the central star, planet formation, ejection by jets, and photoevaporation by ultraviolet radiation from the central star and nearby stars. As a result, the young star becomes a weakly lined T Tauri star, which, over hundreds of millions of years, evolves into an ordinary Sun-like star, dependent on its initial mass.

Accretion of Planets

Artist's impression of a protoplanetary disk showing a young star at its center

Self-accretion of cosmic dust accelerates the growth of the particles into boulder-sized planetesimals. The more massive planetesimals accrete some smaller ones, while others shatter in collisions. Accretion disks are common around smaller stars, or stellar remnants in a close binary, or black holes surrounded by material, such as those at the centers of galaxies. Some dynamics in the disk, such as dynamical friction, are necessary to allow orbiting gas to lose angular momentum and fall onto the central massive object. Occasionally, this can result in stellar surface fusion.

In the formation of terrestrial planets or planetary cores, several stages can be considered. First, when gas and dust grains collide, they agglomerate by microphysical processes like van der Waals forces and electromagnetic forces, forming micrometer-sized particles; during this stage, accumulation mechanisms are largely non-gravitational in nature. However, planetesimal formation in the centimeter-to-meter range is not well understood, and no convincing explanation is offered as to why such grains would accumulate rather than simply rebound. In particular, it is still not clear how these objects grow to become 0.1–1 km (0.06–0.6 mi) sized planetesimals; this problem is known as the "meter size barrier": As dust particles grow by coagulation, they acquire increasingly large relative velocities with respect to other particles in their vicinity, as well as a systematic inward drift velocity, that leads to destructive collisions, and thereby limit the growth of the aggregates to some maximum size. Ward (1996) suggests that when slow moving grains collide, the very low, yet non-zero, gravity of colliding grains impedes their escape. It is also thought that grain fragmentation plays an important role replenishing small grains and keeping the disk thick, but also in maintaining a relatively high abundance of solids of all sizes.

A number of mechanisms have been proposed for crossing the 'meter-sized' barrier. Local concentrations of pebbles may form, which then gravitationally collapse into planetesimals the size of large asteroids. These concentrations can occur passively due to the structure of the gas disk, for example, between eddies, at pressure bumps, at the edge of a gap created by a giant planet, or at the boundaries of turbulent regions of the disk. Or, the particles may take an active role in their concentration via a feedback mechanism referred to as a streaming instability. In a streaming instability the interaction between the solids and the gas in the protoplanetary disk results in the growth of local concentrations, as new particles accumulate in the wake of small concentrations, causing them to grow into massive filaments. Alternatively, if the grains that form due to the agglomeration of dust are highly porous their growth may continue until they become large enough to collapse due to their own gravity. The low density of these objects allows them to remain strongly coupled with the gas, thereby avoiding high velocity collisions which could result in their erosion or fragmentation.

Grains eventually stick together to form mountain-size (or larger) bodies called planetesimals. Collisions and gravitational interactions between planetesimals combine to produce Moon-size planetary embryos (protoplanets) over roughly 0.1–1 million years. Finally, the planetary embryos collide to form planets over 10–100 million years. The planetesimals are massive enough that mutual gravitational interactions are significant enough to be taken into account when computing their evolution. Growth is aided by orbital decay of smaller bodies due to gas drag, which prevents them from being stranded between orbits of the embryos. Further collisions and accumulation lead to terrestrial planets or the core of giant planets.

If the planetesimals formed via the gravitational collapse of local concentrations of pebbles their growth into planetary embryos and the cores of giant planets is dominated by the further accretions of pebbles. Pebble accretion is aided by the gas drag felt by objects as they accelerate toward a massive body. Gas drag slows the pebbles below the escape velocity of the massive body causing them to spiral toward and to be accreted by it. Pebble accretion may accelerate the formation of planets by a factor of 1000 compared to the accretion of planetesimals, allowing giant planets to form before the dissipation of the gas disk. Yet, core growth via pebble accretion appears incompatible with the final masses and compositions of Uranus and Neptune.

The formation of terrestrial planets differs from that of giant gas planets, also called Jovian planets. The particles that make up the terrestrial planets are made from metal and rock that condense in the inner Solar System. However, Jovian planets begin as large, icy planetesimals, which then capture hydrogen and helium gas from the solar nebula. Differentiation between these two classes of planetesimals arise due to the frost line of the solar nebula.

Accretion of Asteroids

Meteorites contain a record of accretion and impacts during all stages of asteroid origin and evolution; however, the mechanism of asteroid accretion and growth is not well understood. Evidence suggests the main growth of asteroids can result from gas-assisted accretion of chondrules, which are millimeter-sized spherules that form as molten (or partially molten) droplets in space before being accreted to their parent asteroids. In the inner Solar System, chondrules appear to have been crucial for initiating accretion. The tiny mass of asteroids may be partly due to inefficient chondrule formation beyond 2 AU, or less-efficient delivery of chondrules from near the protostar. Also, impacts controlled the formation and destruction of asteroids, and are thought to be a major factor in their geological evolution.

Chondrules in a chondrite meteorite. A millimeter scale is shown.

Chondrules, metal grains, and other components likely formed in the solar nebula. These accreted together to form parent asteroids. Some of these bodies subsequently melted, forming metallic cores and olivine-rich mantles; others were aqueously altered. After the asteroids had cooled, they were eroded by impacts for 4.5 billion years, or disrupted.

For accretion to occur, impact velocities must be less than about twice the escape velocity, which is about 140 m/s (460 ft/s) for a 100 km (60 mi) radius asteroid. Simple models for accretion in the asteroid belt generally assume micrometer-sized dust grains sticking together and settling to the midplane of the nebula to form a dense layer of dust, which, because of gravitational forces, was converted into a disk of kilometer-sized planetesimals. But, several arguments suggest that asteroids may not have accreted this way.

Accretion of Comets

The Helix Nebula has a cometary Oort cloud

Comets, or their precursors, formed in the outer Solar System, possibly millions of years before planet formation. How and when comets formed is debated, with distinct implications for Solar System formation, dynamics, and geology. Three-dimensional computer simulations indicate the major structural features observed on cometary nuclei can be explained by pairwise low velocity accretion of weak cometesimals. The currently favored formation mechanism is that of the nebular hypothesis, which states that comets are probably a remnant of the original planetesimal "building blocks" from which the planets grew.

Astronomers think that comets originate in both the Oort cloud and the scattered disk. The scattered disk was created when Neptune migrated outward into the proto-Kuiper belt, which at the time was much closer to the Sun, and left in its wake a population of dynamically stable objects that could never be affected by its orbit (the Kuiper belt proper), and a population whose perihelia are close enough that Neptune can still disturb them as it travels around the Sun (the scattered disk). Because the scattered disk is dynamically active and the Kuiper belt relatively dynamically stable, the scattered disk is now seen as the most likely point of origin for periodic comets. The classic Oort cloud theory states that the Oort cloud, a sphere measuring about 50,000 AU (0.24 pc) in radius, formed at the same time as the solar nebula and occasionally releases comets into the inner Solar System as a giant planet or star passes nearby and causes gravitational disruptions. Examples of such comet clouds may already have been seen in the Helix Nebula.

The *Rosetta* mission to comet 67P/Churyumov–Gerasimenko determined in 2015 that when Sun's heat penetrates the surface, it triggers evaporation (sublimation) of buried ice. While some of the resulting water vapour may escape from the nucleus, 80% of it recondenses in layers beneath the surface. This observation implies that the thin ice-rich layers exposed close to the surface may be a consequence of cometary activity and evolution, and that global layering does not necessarily occur early in the comet's formation history. While most scientists thought that all the evidence indicated that the structure of nuclei of comets is processed rubble piles of smaller ice planetesimals of a previous generation, the *Rosetta* mission dispelled the idea that comets are "rubble piles" of disparate material.

Chandrasekhar Limit

The Chandrasekhar limit is the maximum mass of a stable white dwarf star. The limit was first indicated in papers published by Wilhelm Anderson and E. C. Stoner, and was named after Subrahmanyan Chandrasekhar, the Indian astrophysicist who independently discovered and improved upon the accuracy of the calculation in 1930, at the age of 19, in India. This limit was initially ignored by the community of scientists because such a limit would logically require the existence of black holes, which were considered a scientific impossibility at the time. White dwarfs resist gravitational collapse primarily through electron degeneracy pressure. (By comparison, main sequence stars resist collapse through thermal pressure.) The Chandrasekhar limit is the mass above which electron degeneracy pressure in the star's core is insufficient to balance the star's own gravitational self-attraction. Consequently, white dwarfs with masses greater than the limit would be subject to further gravitational collapse, evolving into a different type of stellar remnant, such as a neutron star or black hole. (However, white dwarfs generally avoid this fate by exploding before they undergo collapse.) Those with masses under the limit remain stable as white dwarfs.

The currently accepted value of the limit is about 1.39 $M_\odot$ (2.765 × 10^{30} kg).

Physics

Electron degeneracy pressure is a quantum-mechanical effect arising from the Pauli exclusion principle. Since electrons are fermions, no two electrons can be in the same state, so not all electrons can be in the minimum-energy level. Rather, electrons must occupy a band of energy levels. Compression of the electron gas increases the number of electrons in a given volume and raises the maximum energy level in the occupied band. Therefore, the energy of the electrons will increase upon compression, so pressure must be exerted on the electron gas to compress it, producing electron degeneracy pressure. With sufficient compression, electrons are forced into nuclei in the process of electron capture, relieving the pressure.

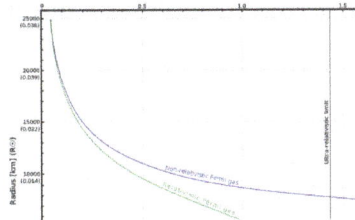

Radius–mass relations for a model white dwarf. The green curve uses the general pressure law for an ideal Fermi gas, while the blue curve is for a non-relativistic ideal Fermi gas. The black line marks the ultrarelativistic limit.

In the nonrelativistic case, electron degeneracy pressure gives rise to an equation of state of the form $P = K_1 \rho^{\frac{5}{3}}$, where P is the pressure, ρ is the mass density, and K_1 is a constant. Solving the hydrostatic equation then leads to a model white dwarf which is a polytrope of index 3/2 and therefore has radius inversely proportional to the cube root of its mass, and volume inversely proportional to its mass.

As the mass of a model white dwarf increases, the typical energies to which degeneracy pressure forces the electrons are no longer negligible relative to their rest masses. The velocities of the electrons approach the speed of light, and special relativity must be taken into account. In the strongly relativistic limit, the equation of state takes the form $P = K_2 \rho^{\frac{4}{3}}$. This will yield a polytrope of index 3, which will have a total mass, M_{limit} say, depending only on K_2.

For a fully relativistic treatment, the equation of state used will interpolate between the equations $P = K_1 \rho^{\frac{5}{3}}$ for small ρ and $P = K_2 \rho^{\frac{4}{3}}$ for large ρ. When this is done, the model radius still decreases with mass, but becomes zero at M_{limit}. This is the Chandrasekhar limit. The curves of radius against mass for the non-relativistic and relativistic models are shown in the graph. They are colored blue and green, respectively. μ_e has been set equal to 2. Radius is measured in standard solar radii or kilometers, and mass in standard solar masses.

Calculated values for the limit will vary depending on the nuclear composition of the mass. Chandrasekhar gives the following expression, based on the equation of state for an ideal Fermi gas:

$$M_{\text{limit}} = \frac{\omega_3^0 \sqrt{3\pi}}{2} \left(\frac{\hbar c}{G} \right)^{3/2} \frac{1}{(\mu_e m_H)^2},$$

where:

- $\hbar$ is the reduced Planck constant

- c is the speed of light

- G is the gravitational constant

- μ_e is the average molecular weight per electron, which depends upon the chemical composition of the star.

- m_H is the mass of the hydrogen atom.

- $\omega_3^0 \approx 2.018236$ is a constant connected with the solution to the Lane-Emden equation.

As $\sqrt{\hbar c / G}$ is the Planck mass, the limit is of the order of

$$\frac{MPl^3}{m_H^2}$$

A more accurate value of the limit than that given by this simple model requires adjusting for various factors, including electrostatic interactions between the electrons and nuclei and effects caused by nonzero temperature. Lieb and Yau have given a rigorous derivation of the limit from a relativistic many-particle Schrödinger equation.

History

In 1926, the British physicist Ralph H. Fowler observed that the relationship among the density, energy and temperature of white dwarfs could be explained by viewing them as a gas of nonrelativistic, non-interacting electrons and nuclei which obeyed Fermi–Dirac statistics. This Fermi gas model was then used by the British physicist Edmund Clifton Stoner in 1929 to calculate the relationship among the mass, radius, and density of white dwarfs, assuming them to be homogeneous spheres. Wilhelm Anderson applied a relativistic correction to this model, giving rise to a maximum possible mass of approximately 1.37×10^{30} kg. In 1930, Stoner derived the internal energy–density equation of state for a Fermi gas, and was then able to treat the mass–radius relationship in a fully relativistic manner, giving a limiting mass of approximately (for μ_e=2.5) $2.19 \cdot 10^{30}$ kg. Stoner went on to derive the pressure–density equation of state, which he published in 1932. These equations of state were also previously published by the Soviet physicist Yakov Frenkel in 1928, together with some other remarks on the physics of degenerate matter. Frenkel's work, however, was ignored by the astronomical and astrophysical community.

A series of papers published between 1931 and 1935 had its beginning on a trip from India to England in 1930, where the Indian physicist Subrahmanyan Chandrasekhar worked on the calculation of the statistics of a degenerate Fermi gas. In these papers, Chandrasekhar solved the hydrostatic equation together with the nonrelativistic Fermi gas equation of state, and also treated the case of a relativistic Fermi gas, giving rise to the value of the limit shown above. Chandrasekhar reviews this work in his Nobel Prize lecture. This value was also computed in 1932 by the Soviet physicist Lev Davidovich Landau, who, however, did not apply it to white dwarfs.

Chandrasekhar's work on the limit aroused controversy, owing to the opposition of the British astrophysicist Arthur Eddington. Eddington was aware that the existence of black holes was theoretically possible, and also realized that the existence of the limit made their formation possible. However, he was unwilling to accept that this could happen. After a talk by Chandrasekhar on the limit in 1935, he replied:

The star has to go on radiating and radiating and contracting and contracting until, I suppose, it gets down to a few km radius, when gravity becomes strong enough to hold in the radiation, and the star can at last find peace. ... I think there should be a law of Nature to prevent a star from behaving in this absurd way!

Eddington's proposed solution to the perceived problem was to modify relativistic mechanics so as to make the law $P = K_1 \rho^{5/3}$ universally applicable, even for large ρ. Although Niels Bohr, Fowler, Wolfgang Pauli, and other physicists agreed with Chandrasekhar's analysis, at the time, owing to Eddington's status, they were unwilling to publicly support Chandrasekhar. Through the rest of his life, Eddington held to his position in his writings, including his work on his fundamental theory. The drama associated with this disagreement is one of the main themes of *Empire of the Stars*, Arthur I. Miller's biography of Chandrasekhar. In Miller's view:

Chandra's discovery might well have transformed and accelerated developments in both physics and astrophysics in the 1930s. Instead, Eddington's heavy-handed intervention lent weighty support to the conservative community astrophysicists, who steadfastly refused even to consider the idea that stars might collapse to nothing. As a result, Chandra's work was almost forgotten.

Applications

The core of a star is kept from collapsing by the heat generated by the fusion of nuclei of lighter elements into heavier ones. At various stages of stellar evolution, the nuclei required for this process will be exhausted, and the core will collapse, causing it to become denser and hotter. A critical situation arises when iron accumulates in the core, since iron nuclei are incapable of generating further energy through fusion. If the core becomes sufficiently dense, electron degeneracy pressure will play a significant part in stabilizing it against gravitational collapse.

If a main-sequence star is not too massive (less than approximately 8 solar masses), it will eventually shed enough mass to form a white dwarf having mass below the Chandrasekhar limit, which will consist of the former core of the star. For more-massive stars, electron degeneracy pressure will not keep the iron core from collapsing to very great density, leading to formation of a neutron star, black hole, or, speculatively, a quark star. (For very massive, low-metallicity stars, it is also possible that instabilities will destroy the star completely.) During the collapse, neutrons are formed by the capture of electrons by protons in the process of electron capture, leading to the emission of neutrinos. The decrease in gravitational potential energy of the collapsing core releases a large amount of energy which is on the order of 10^{46} joules (100 foes). Most of this energy is carried away by the emitted neutrinos. This process is believed to be responsible for supernovae of types Ib, Ic, and II.

TYPE IA SUPERNOVAE derive their energy from runaway fusion of the nuclei in the interior of a white dwarf. This fate may befall carbon–oxygen white dwarfs that accrete matter from a companion giant star, leading to a steadily increasing mass. As the white dwarf's mass approaches the

Chandrasekhar limit, its central density increases, and, as a result of compressional heating, its temperature also increases. This eventually ignites nuclear fusion reactions, leading to an immediate carbon detonation which disrupts the star and causes the supernova.

A strong indication of the reliability of Chandrasekhar's formula is that the absolute magnitudes of supernovae of Type Ia are all approximately the same; at maximum luminosity, M_V is approximately -19.3, with a standard deviation of no more than 0.3.[1] A 1-sigma interval therefore represents a factor of less than 2 in luminosity. This seems to indicate that all type Ia supernovae convert approximately the same amount of mass to energy.

Super-chandrasekhar Mass Supernovae

In April 2003, the Supernova Legacy Survey observed a type Ia supernova, designated SN-LS-03D3bb, in a galaxy approximately 4 billion light years away. According to a group of astronomers at the University of Toronto and elsewhere, the observations of this supernova are best explained by assuming that it arose from a white dwarf which grew to twice the mass of the Sun before exploding. They believe that the star, dubbed the "Champagne Supernova" by University of Oklahoma astronomer David R. Branch, may have been spinning so fast that a centrifugal tendency allowed it to exceed the limit. Alternatively, the supernova may have resulted from the merger of two white dwarfs, so that the limit was only violated momentarily. Nevertheless, they point out that this observation poses a challenge to the use of type Ia supernovae as standard candles.

Since the observation of the Champagne Supernova in 2003, more very bright type Ia supernovae have been observed that are thought to have originated from white dwarfs whose masses exceeded the Chandrasekhar limit. These include SN 2006gz, SN 2007if and SN 2009dc. The super-Chandrasekhar mass white dwarfs that gave rise to these supernovae are believed to have had masses up to 2.4–2.8 solar masses. One way to potentially explain the problem of the Champagne Supernova was considering it the result of an aspherical explosion of a white dwarf. However, spectropolarimetric observations of SN 2009dc showed it had a polarization smaller than 0.3, making the large asphericity theory unlikely.

Tolman–Oppenheimer–Volkoff Limit

After a supernova explosion, a neutron star may be left behind. Like white dwarfs these objects are extremely compact and are supported by degeneracy pressure, but a neutron star is so massive and compressed that electrons and protons have combined to form neutrons, and the star is thus supported by neutron degeneracy pressure instead of electron degeneracy pressure. The limit of neutron degeneracy pressure, analogous to the Chandrasekhar limit, is known as the Tolman–Oppenheimer–Volkoff limit.

Eddington Number

In astrophysics, the Eddington number, N_{Edd}, is the number of protons in the observable universe. The term honors the British astrophysicist Arthur Eddington, who in 1938 was the first to propose a value of N_{Edd} and to explain why this number might be important for cosmology and the foundations of physics.

Arthur Stanley Eddington (1882–1944)

History

Eddington argued that the value of the fine-structure constant, α, could be obtained by pure deduction. He related α to the Eddington number, which was his estimate of the number of protons in the universe. This led him in 1929 to conjecture that α was exactly 1/137. Other physicists did not adopt this conjecture and did not accept his argument.

In the late 1930s, the best experimental value of the fine-structure constant, α, was approximately 1/136. Eddington then argued, from aesthetic and numerological considerations, that α should be exactly 1/136. He devised a "proof" that $N_{Edd} = 136 \times 2^{256}$, or about 1.57×10^{79}. Some estimates of N_{Edd} point to a value of about 10^{80}. These estimates assume that all matter can be taken to be hydrogen and require assumed values for the number and size of galaxies and stars in the universe.

Attempts to find a mathematical basis for this dimensionless constant have continued up to the present time.

In the 1938 Tarner Lecture at Trinity College, Cambridge, Eddington averred that:

I believe there are 15 747 724 136 275 002 577 605 653 961 181 555 468 044 717 914 527 116 709 366 231 425 076 185 631 031 296 protons in the universe and the same number of electrons.

This large number was soon named the "Eddington number."

Shortly thereafter, improved measurements of α yielded values closer to 1/137, whereupon Eddington changed his "proof" to show that α had to be exactly 1/137.

Recent Theory

The most precise value of α (obtained experimentally in 2012) is:

$$\alpha^{-1} = 137.035999174(35).$$

Consequently, no one maintains any longer that α is the reciprocal of an integer. Nor does anyone take seriously a mathematical relationship between α and N_{Edd}.

On possible roles for N_{Edd} in contemporary cosmology, especially its connection with large number coincidences.

Hayashi Track

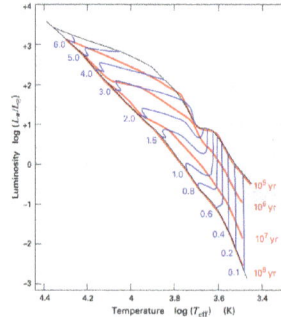

Stellar evolution tracks (blue lines) for the pre-main-sequence. The nearly vertical curves are Hayashi tracks.

Low-mass stars have nearly vertical evolution tracks until they arrive on the main sequence. For more-massive stars, the Hayashi track bends to the left into the Henyey track. Even more-massive stars are born directly onto the Henyey track.

The end (leftmost point) of every track is labeled with the star's mass in solar masses ($M_{\square}$), and represents its position on the main sequence. The red curves labeled in years are isochrones at the given ages. In other words, stars 10^5 years old lie along the curve labeled 10^5, and similarly for the other 3 isochrones.

The Hayashi track is a luminosity–temperature relationship obeyed by infant stars of less than $3\,M_{\odot}$ in the pre-main-sequence phase (PMS phase) of stellar evolution. It is named after Japanese astrophysicist Chushiro Hayashi. On the Hertzsprung–Russell diagram, which plots luminosity against temperature, the track is a nearly vertical curve. After a protostar ends its phase of rapid contraction and becomes a T Tauri star, it is extremely luminous. The star continues to contract, but much more slowly. While slowly contracting, the star follows the Hayashi track downwards, becoming several times less luminous but staying at roughly the same surface temperature, until either a radiative zone develops, at which point the star starts following the Henyey track, or nuclear fusion begins, marking its entry onto the main sequence.

The shape and position of the Hayashi track on the Hertzsprung–Russell diagram depends on the star's mass and chemical composition. For solar-mass stars, the track lies at a temperature of roughly 4000 K. Stars on the track are nearly fully convective and have their opacity dominated by hydrogen ions. Stars less than $0.5\,M_{\odot}$ are fully convective even on the main sequence, but their opacity begins to be dominated by Kramers' opacity law after nuclear fusion begins, thus moving them off the Hayashi track. Stars between 0.5 and $3\,M_{\odot}$ develop a radiative zone prior to reaching the main sequence. Stars between 3 and $10\,M_{\odot}$ are fully radiative at the beginning of the pre-main-sequence. Even heavier stars are born onto the main sequence, with no PMS evolution.

At an end of a low- or intermediate-mass star's life, the star follows an analogue of the Hayashi track, but in reverse—it increases in luminosity, expands, and stays at roughly the same temperature, eventually becoming a red giant.

History

In 1961, Professor Chushiro Hayashi published two papers that led to the concept of the pre-main-sequence and form the basis of the modern understanding of early stellar evolution. Hayashi realized that the existing model, in which stars are assumed to be in radiative equilibrium with no substantial convection zone, cannot explain the shape of the red giant branch. He therefore replaced the model by including the effects of thick convection zones on a star's interior.

A few years prior, Osterbrock proposed deep convection zones with efficient convection, analyzing them using the opacity of H- ions (the dominant opacity source in cool atmospheres) in temperatures below 5000K. However, the earliest numerical models of Sun-like stars did not follow up on this work and continued to assume radiative equilibrium.

In his 1961 papers, Hayashi showed that the convective envelope of a star is determined by:

$$E = 4\pi G^{3/2} (\mu H / k)^{5/2} M^{1/2} R^{3/2} P / T^{5/2}$$

where E is unitless, and not the energy. Modelling stars as polytropes with index 3/2--in other words, assuming they follow a pressure-density relationship of $P = K\rho^{5/3}$ —he found that E=45 is the maximum for a quasistatic star. If a star is not contracting rapidly, E=45 defines a curve on the HR diagram, to the right of which the star cannot exist. He then computed the evolutionary tracks and isochrones (luminosity-temperature distributions of stars at a given age) for a variety of stellar masses and noted that NGC2264, a very young star cluster, fits the isochrones well. In particular, he calculated much lower ages for solar-type stars in NGC2264 and predicted that these stars were rapidly contracting T Tauri stars.

In 1962, Hayashi published a 183-page review of stellar evolution. Here, he discussed the evolution of stars born in the forbidden region. These stars rapidly contract due to gravity before settling to a quasistatic, fully convective state on the Hayashi tracks.

In 1965, numerical models by Iben and Ezer & Cameron realistically simulated pre-main-sequence evolution, including the Henyey track that stars follow after leaving the Hayashi track. These standard PMS tracks can still be found in textbooks on stellar evolution.

Forbidden Zone and Hayashi Limit

The forbidden zone is the region on the HR diagram to the right of the Hayashi track where no star in hydrostatic equilibrium, even those that are partially or fully radiative, can be. Newborn protostars start out in this zone, but are not in hydrostatic equilibrium and will rapidly move towards the Hayashi track.

Because stars emit light via blackbody radiation, the power per unit surface area they emit is given by the Stefan-Boltzmann law:

$$j^\star = \sigma T^4$$

The star's luminosity is therefore given by:

$$L = 4\pi R^2 \sigma T^4$$

For a given L, a lower temperature implies a larger radius, and vice versa. Thus, the Hayashi track separates the HR diagram into two regions: the allowed region to the left, with high temperatures and smaller radii for each luminosity, and the forbidden region to the right, with lower temperatures and correspondingly higher radii. The Hayashi limit can refer to either the lower bound in temperature or the upper bound on radius defined by the Hayashi track.

The region to the right is forbidden because it can be shown that a star in the region must have a temperature gradient of:

$$\frac{\ln T}{d \ln P} > 0.4$$

where $\frac{d \ln T}{d \ln P} = 0.4$ for a monatomic ideal gas undergoing adiabatic expansion or contraction. A temperature gradient greater than 0.4 is therefore called superadiabatic.

Consider a star with a superadiabatic gradient. Imagine a parcel of gas that starts at radial position r, but moves upwards to r+dr in a sufficiently short time that it exchanges negligible heat with its surroundings—in other words, the process is adiabatic. The pressure of the surroundings, as well as that of the parcel, decreases by some amount dP. The parcel's temperature changes by

$dT = 0.4 \frac{T}{P} dP.$ The temperature of the surroundings also decreases, but by some amount dT' that is greater than dT. The parcel therefore ends up being hotter than its surroundings. Since the ideal

gas law can be written $P = \frac{\rho RT}{\mu}$, a higher temperature implies a lower density at the same pres-

sure. The parcel is therefore also less dense than its surroundings. This will cause it to rise even more, and the parcel will become even less dense than its new surroundings.

Clearly, this situation is not stable. In fact, a superadiabatic gradient causes convection. Convection tends to lower the temperature gradient because the rising parcel of gas will eventually be dispersed, dumping its excess thermal and kinetic energy into its surroundings and heating up said surroundings. In stars, the convection process is known to be highly efficient, with a typical

$\frac{d \ln T}{d \ln P}$ that only exceeds the adiabatic gradient by 1 part in 10 million.

If a star is placed in the forbidden zone, with a temperature gradient much greater than 0.4, it will experience rapid convection that brings the gradient down. Since this convection will drastically change the star's pressure and temperature distribution, the star is not in hydrostatic equilibrium, and will contract until it is.

A star far to the left of the Hayashi track has a temperature gradient smaller than adiabatic. This means that if a parcel of gas rises a tiny bit, it will be more dense than its surroundings and sink back to where it came from. Convection therefore does not occur, and almost all energy output is carried radiatively.

Star Formation

Stars form when small regions of a giant molecular cloud collapse under their own gravity, becoming protostars. The collapse releases gravitational energy, which heats up the protostar. This process occurs on the free-fall timescale, which is roughly 100,000 years for solar-mass protostars, and ends when the protostar reaches approximately 4000 K. This is known as the Hayashi boundary, and at this point, the protostar is on the Hayashi track. At this point, they are known as T Tauri stars and continue to contract, but much more slowly. As they contract, they decrease in luminosity because less surface area becomes available for emitting light. The Hayashi track gives the resulting change in temperature, which will be minimal compared to the change in luminosity because the Hayashi track is nearly vertical. In other words, on the HR diagram, a T Tauri star starts out on the Hayashi track with a high luminosity and moves downward along the track as time passes.

The Hayashi track describes a fully convective star. This is a good approximation for very young pre-main-sequence stars they are still cool and highly opaque, so that radiative transport is insufficient to carry away the generated energy and convection must occur. Stars less massive than $0.5\,M_{\odot}$ remain fully convective, and therefore remain on the Hayashi track, throughout their pre-main-sequence stage, joining the main sequence at the bottom of the Hayashi track. Stars heavier than $0.5\,M_{\odot}$ have higher interior temperatures, which decreases their central opacity and allows radiation to carry away large amounts of energy. This allows a radiative zone to develop around the star's core. The star is then no longer on the Hayashi track, and experiences a period of rapidly increasing temperature at nearly constant luminosity. This is called the Henyey track, and ends when temperatures are high enough to ignite hydrogen fusion in the core. The star is then on the main sequence.

Lower-mass stars follow the Hayashi track until the track intersects with the main sequence, at which point hydrogen fusion begins and the star follows the main sequence. Even lower-mass 'stars' never achieve the conditions necessary to fuse hydrogen and become brown dwarfs.

Derivation

The exact shape and position of the Hayashi track can only be computed numerically using computer models. Nevertheless, we can make an extremely crude analytical argument that captures most of the track's properties. The following derivation loosely follows that of Kippenhahn, Weigert, and Weiss in *Stellar Structure and Evolution*.

In our simple model, a star is assumed to consist of a fully convective interior inside of a fully radiative atmosphere.

The convective interior is assumed to be an ideal monatomic gas with a perfectly adiabatic temperature gradient:

$$\frac{d\ln T}{d\ln P} = 0.4$$

This quantity is sometimes labelled ∇. The following adiabatic equation therefore holds true for the entire interior:

$$P^{1-\gamma}T^{\gamma} = C$$

where γ is the adiabatic gamma, which is 5/3 for an ideal monatomic gas. The ideal gas law says:

$$P = NkT / V$$

$$= \frac{\rho kT}{\mu H}$$

$$= (\frac{k\rho C}{\mu H})^{\gamma}$$

where μ is the molecular weight per particle and H is (to a very good approximation) the mass of a hydrogen atom. This equation represents a polytrope of index 1.5, since a polytrope is defined by $P = K\rho^{1+1/n}$, where n=1.5 is the polytropic index. Applying the equation to the center of the star

gives: $P_c = (\frac{k\rho_c C}{\mu H})^{\gamma}$ We can solve for C:

$$C = \frac{\mu H P_c^{1/\gamma}}{\rho_c k}$$

But for any polytrope, $P_c = W_n \frac{GM^2}{R^4}$, $\rho_c = K_n \rho_{avg}$, and $R^{\frac{3-n}{n}} M^{\frac{n-1}{n}} = \frac{K}{GN_n}$. W_n, K_n, N_n and K are all constants independent of pressure and density, and the average density is defined as $\rho_{avg} \equiv \frac{M}{4/3\pi R^3}$. Plugging all 3 equations into the equation for C, we have:

$$C \sim M^{2-\gamma} R^{3\gamma-4}$$

where all multiplicative constants have been ignored. Recall that our original definition of C was:

$$P^{1-\gamma} T^{\gamma} = C$$

We therefore have, for any star of mass M and radius R:

$$P^{1-\gamma} T^{\gamma} \sim M^{2-\gamma} R^{3\gamma-4}$$

$$\ln P = \frac{2-\gamma}{1-\gamma} \ln M + \frac{3\gamma-4}{1-\gamma} \ln R - \gamma \ln T \qquad (1)$$

We need another relationship between P, T, M, and R, in order to eliminate P. This relationship will come from the atmosphere model.

The atmosphere is assumed to be thin, with average opacity k. Opacity is defined to be optical depth divided by density. Thus, by definition, the optical depth of the stellar surface, also called the photosphere, is:

$$\frac{d\tau}{dr} = k\rho$$

$$\tau = \int_R^\infty k\rho dr$$

$$= k\int_R^\infty \rho dr$$

where R is the stellar radius, also known as the position of the photosphere. The pressure at the surface is:

$$P_0 = \int_R^\infty g\rho dr$$

$$= \frac{GM}{R^2}\int_R^\infty \rho dr$$

$$= \frac{GM\tau}{kR^2}$$

The optical depth at the photosphere turns out to be $\tau = 2/3$. By definition, the temperature of the photosphere is $T = T_{eff}$ where effective temperature is given by $L = 4\pi R^2 T_{eff}^4$. Therefore, the pressure is:

$$P_0 = \frac{GM}{R^2}\frac{2\tau}{3k}$$

We can approximate the opacity to be:

2where a=1, b=3. Plugging this into the pressure equation, we get:

$$P_0 = const(\frac{M}{R^2 T_{eff}^b})^{\frac{1}{a+1}}$$

$$\ln P_0 = \ln const + \frac{1}{a+1}(\ln M - 2\ln R - b\ln T_{eff}) \qquad (2)$$

Finally, we need to eliminate R and introduce L, the luminosity. This can be done with the equation:

$$L = 4\pi R^2 \sigma T_{eff}^4$$

$$\ln R = 0.5\ln L - 2\ln T_{eff} + const \qquad (3)$$

Equation 1 and 2 can now be combined by setting $T = T_{eff}$ and $P = P_0$ in Equation 1, then eliminating P_0. R can be eliminated using Equation 3. After some algebra, and after setting $\gamma = 5/3$, we get:

$$\ln T_{eff} = A \ln L + B \ln M + const$$

where

$$A = \frac{0.75a - 0.25}{5.5a + b + 1.5}$$

$$B = \frac{0.5a + 1.5}{5.5a + b + 1.5}$$

In cool stellar atmospheres (T < 5000 K) like those of newborn stars, the dominant source of opacity is the H- ion, for which $a \approx 1$ and $b \approx 3$, we get $A = 0.05$ and $B = 0.2$.

Since A is much smaller than 1, the Hayashi track is extremely steep: if the luminosity changes by a factor of 2, the temperature only changes by 4 percent. The fact that B is positive indicates that the Hayashi track shifts left on the HR diagram, towards higher temperatures, as mass increases. Although this model is extremely crude, these qualitative observations are fully supported by numerical simulations.

At high temperatures, the atmosphere's opacity begins to be dominated by Kramers' opacity law instead of the H- ion, with a=1 and b=-4.5 In that case, A=0.2 in our crude model, far higher than 0.05, and the star is no longer on the Hayashi track.

In *Stellar Interiors*, Hansen, Kawaler, and Trimble go through a similar derivation without neglecting multiplicative constants, and arrived at:

$$T_{eff} = (2600K)\mu^{13/51}(\frac{M}{M_\odot})^{7/51}(\frac{L}{L_\odot})^{1/102}$$

where μ is the molecular weight per particle. The authors note that the coefficient of 2600K is too low—it should be around 4000K—but this equation nevertheless shows that temperature is nearly independent of luminosity.

Numerical Results

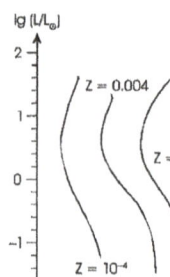

Hayashi tracks of a 0.8 $M_\odot$ star with helium mass fraction 0.245, for 3 different metallicities

The diagram at the top of this article shows numerically computed stellar evolution tracks for various masses. The vertical portions of each track is the Hayashi track. The endpoints of each track lie on the main sequence. The horizontal segments for higher-mass stars show the Henyey track.

It is approximately true that:

$$\frac{\partial \ln T_{eff}}{\partial \ln M} \approx 0.1.$$

The diagram to the right shows how Hayashi tracks change with changes in chemical composition. Z is the star's metallicity, the mass fraction not accounted for by hydrogen or helium. For any given hydrogen mass fraction, increasing Z leads to increasing molecular weight. The dependence of temperature on molecular weight is extremely steep—it is approximately

$$\frac{\partial \ln_{eff}}{\partial \ln} \approx -26..$$

Decreasing Z by a factor of 10 shifts the track right, changing $\ln T_{eff}$ by about 0.05.

Chemical composition affects the Hayashi track in a few ways. The track depends strongly on the atmosphere's opacity, and this opacity is dominated by the H- ion. The abundance of the H- ion is proportional to the density of free electrons, which, in turn, is higher if there are more metals because metals are easier to ionize than hydrogen or helium.

Observational Status

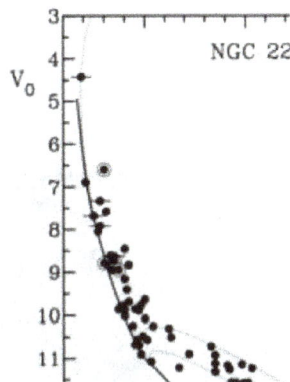

The young star cluster NGC 2264, with a large number of T Tauri stars contracting towards the main sequence. The solid line represents the main sequence, while the two lines above that are the $10^{6.5}$ yr (upper) and $10^{6.7}$ yr (lower) isochrones.

Observational evidence of the Hayashi track comes from color-magnitude plots—the observational equivalent of HR diagrams—of young star clusters. For Hayashi, NGC 2264 provided the first evidence of a population of contracting stars. In 2012, data from NGC 2264 was re-analyzed to account for dust reddening and extinction. The resulting color-magnitude plot is shown at right.

In the upper diagram, the isochrones are curves along which stars of a certain age are expected to lie, assuming that all stars evolve along the Hayashi track. An isochrone is created by taking stars of every conceivable mass, evolving them forwards to the same age, and plotting all of them on the color-magnitude diagram. Most of the stars in NGC 2264 are already on the main sequence (black line), but a substantial population lies between the isochrones for 3.2 million and 5 million years, indicating that the cluster is 3.2-5 million years old and a large population of T Tauri stars is still on their respective Hayashi tracks. Similar results have been obtained for NGC 6530, IC 5146, and NGC 6611.

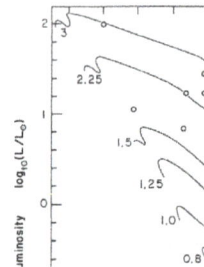

The numbered curves show the Hayashi tracks of stars of that mass (in solar masses). The small circles represent observational data of T Tauri stars. The bold curve to the right is the birthline, above which few stars exist.

The lower diagram shows Hayashi tracks for various masses, along with T Tauri observations collected from a variety of sources. Note the bold curve to the right, representing a stellar birthline. Even though some Hayashi tracks theoretically extend above the birthline, few stars are above it. In effect, stars are 'born' onto the birthline before evolving downwards along their respective Hayashi tracks.

The birthline exists because stars form from overdense cores of giant molecular clouds in an inside-out manner. That is, a small central region first collapses in on itself while the outer shell is still nearly static. The outer envelope then accretes onto the central protostar. Before the accretion is over, the protostar is hidden from view, and therefore not plotted on the color-magnitude diagram. When the envelope finishes accreting, the star is revealed and appears on the birthline.

Hertzsprung–Russell Diagram

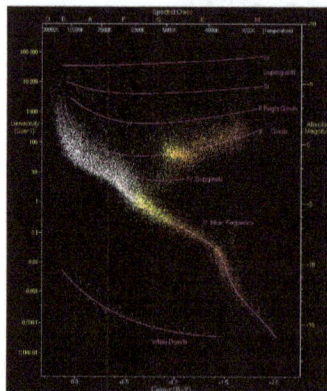

Hertzsprung–Russell diagram with 22,000 stars plotted from the Hipparcos Catalogue and 1,000 from the Gliese Catalogue of nearby stars. Stars tend to fall only into certain regions of the diagram. The most prominent is the diagonal, going from the upper-left (hot and bright) to the lower-right (cooler and less bright), called the main sequence. In the lower-left is where white dwarfs are found, and above the main sequence are the subgiants, giants and supergiants. The Sun is found on the main sequence at luminosity 1 (absolute magnitude 4.8) and B–V color index 0.66 (temperature 5780 K, spectral type G2V).

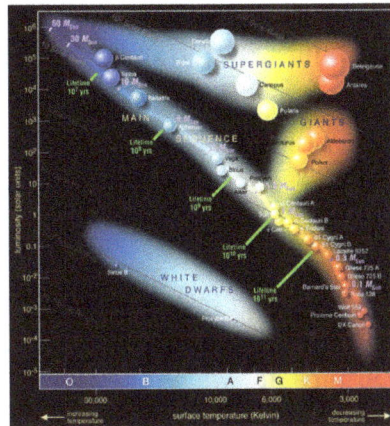

An HR diagram showing many well known stars in the Milky Way galaxy.

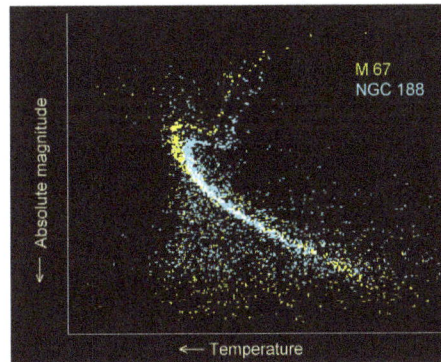

HR diagrams for two open clusters, M67 and NGC 188, showing the main-sequence turn-off at different ages.

The Hertzsprung–Russell diagram, abbreviated H–R diagram or HRD, is a scatter graph of stars showing the relationship between the stars' absolute magnitudes or luminosities versus their spectral classifications or effective temperatures. More simply, it plots each star on a graph measuring the star's brightness against its temperature (color). It does not map any locations of stars.

The diagram was created circa 1910 by Ejnar Hertzsprung and Henry Norris Russell and represents a major step towards an understanding of stellar evolution or "the way in which stars undergo sequences of dynamic and radical changes over time".

Historical Background

In the nineteenth-century large-scale photographic spectroscopic surveys of stars were performed at Harvard College Observatory, producing spectral classifications for tens of thousands of stars, culminating ultimately in the Henry Draper Catalogue. In one segment of this work Antonia Maury included divisions of the stars by the width of their spectral lines. Hertzsprung noted that stars described with narrow lines tended to have smaller proper motions than the others of the same spectral classification. He took this as an indication of greater luminosity for the narrow-line stars, and computed secular parallaxes for several groups of these, allowing him to estimate their absolute magnitude.

In 1910 Hans Rosenberg published a diagram plotting the apparent magnitude of stars in the Pleiades cluster against the strengths of the Calcium K line and two Hydrogen Balmer lines. These spectral lines serve as a proxy for the temperature of the star, an early form of spectral classification. The apparent magnitude of stars in the same cluster is equivalent to their absolute magnitude and so this early diagram was effectively a plot of luminosity against temperature. The same type of diagram is still used today as a means of showing the stars in clusters without having to initially know their distance and luminosity. Hertzsprung had already been working with this type of diagram, but his first publications showing it were not until 1911. This was also the form of the diagram using apparent magnitudes of a cluster of stars all at the same distance.

Russell's early (1913) versions of the diagram included Maury's giant stars identified by Hertzsprung, those nearby stars with parallaxes measured at the time, stars from the Hyades (a nearby open cluster), and several moving groups, for which the moving cluster method could be used to derive distances and thereby obtain absolute magnitudes for those stars.

Forms of Diagram

There are several forms of the Hertzsprung–Russell diagram, and the nomenclature is not very well defined. All forms share the same general layout: stars of greater luminosity are toward the top of the diagram, and stars with higher surface temperature are toward the left side of the diagram.

The original diagram displayed the spectral type of stars on the horizontal axis and the absolute visual magnitude on the vertical axis. The spectral type is not a numerical quantity, but the sequence of spectral types is a monotonic series that reflects the stellar surface temperature. Modern observational versions of the chart replace spectral type by a color index (in diagrams made in the middle of the 20th Century, most often the B-V color) of the stars. This type of diagram is what is often called an observational Hertzsprung–Russell diagram, or specifically a color-magnitude diagram (CMD), and it is often used by observers. In cases where the stars are known to be at identical distances such as within a star cluster, a color-magnitude diagram is often used to describe the stars of the cluster with a plot in which the vertical axis is the apparent magnitude of the stars. For cluster members, by assumption there is a single additive constant difference between their apparent and absolute magnitudes, called the distance modulus, for all of that cluster of stars. Early studies of nearby open clusters (like the Hyades and Pleiades) by Hertzsprung and Rosenberg produced the first CMDs, antedating by a few years Russell's influential synthesis of the diagram collecting data for all stars for which absolute magnitudes could be determined.

Another form of the diagram plots the effective surface temperature of the star on one axis and the luminosity of the star on the other, almost invariably in a log-log plot. Theoretical calculations of stellar structure and the evolution of stars produce plots that match those from observations. This type of diagram could be called *temperature-luminosity diagram*, but this term is hardly ever used; when the distinction is made, this form is called the *theoretical Hertzsprung–Russell diagram* instead. A peculiar characteristic of this form of the H–R diagram is that the temperatures are plotted from high temperature to low temperature, which aids in comparing this form of the H–R diagram with the observational form.

Although the two types of diagrams are similar, astronomers make a sharp distinction between the two. The reason for this distinction is that the exact transformation from one to the other is not trivial. To go between effective temperature and color requires a color-temperature relation, and constructing that is difficult; it is known to be a function of stellar composition and can be affected by other factors like stellar rotation. When converting luminosity or absolute bolometric magnitude to apparent or absolute visual magnitude, one requires a bolometric correction, which may or may not come from the same source as the color-temperature relation. One also needs to know the distance to the observed objects (*i.e.*, the distance modulus) and the effects of interstellar obscuration, both in the color (reddening) and in the apparent magnitude (extinction). For some stars, circumstellar dust also affects colors and apparent brightness. The ideal of direct comparison of theoretical predictions of stellar evolution to observations thus has additional uncertainties incurred in the conversions between theoretical quantities and observations.

Interpretation

Most of the stars occupy the region in the diagram along the line called the main sequence. During the stage of their lives in which stars are found on the main sequence line, they are fusing hydrogen in their cores. The next concentration of stars is on the horizontal branch (helium fusion in the core and hydrogen burning in a shell surrounding the core). Another prominent feature is the Hertzsprung gap located in the region between A5 and G0 spectral type and between +1 and −3 absolute magnitudes (*i.e.* between the top of the main sequence and the giants in the horizontal branch). RR Lyrae variable stars can be found in the left of this gap. Cepheid variables reside in the upper section of the instability strip.

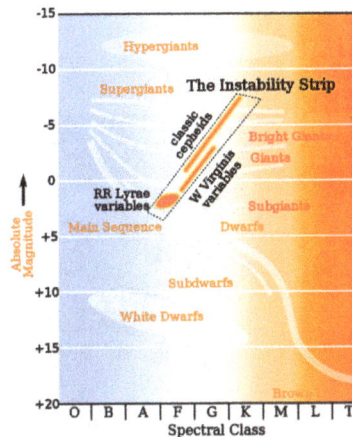

An HR diagram with the instability strip and its components highlighted.

The H-R diagram can be used by scientists to roughly measure how far away a star cluster is from Earth. This can be done by comparing the apparent magnitudes of the stars in the cluster to the absolute magnitudes of stars with known distances (or of model stars). The observed group is then shifted in the vertical direction, until the two main sequences overlap. The difference in magnitude that was bridged in order to match the two groups is called the distance modulus and is a direct measure for the distance (ignoring extinction). This technique is known as main sequence fitting and is a type of spectroscopic parallax.

Diagram's Role in the Development of Stellar Physics

Contemplation of the diagram led astronomers to speculate that it might demonstrate stellar evolution, the main suggestion being that stars collapsed from red giants to dwarf stars, then moving down along the line of the main sequence in the course of their lifetimes. Stars were thought therefore to radiate energy by converting gravitational energy into radiation through the Kelvin–Helmholtz mechanism. This mechanism resulted in an age for the Sun of only tens of millions of years, creating a conflict over the age of the Solar System between astronomers, and biologists and geologists who had evidence that the Earth was far older than that. This conflict was only resolved in the 1930s when nuclear fusion was identified as the source of stellar energy.

However, following Russell's presentation of the diagram to a meeting of the Royal Astronomical Society in 1912, Arthur Eddington was inspired to use it as a basis for developing ideas on stellar physics. In 1926, in his book *The Internal Constitution of the Stars* he explained the physics of how stars fit on the diagram. This was a particularly remarkable development since at that time the major problem of stellar theory, the source of a star's energy, was still unsolved. Thermonuclear energy, and even that stars are largely composed of hydrogen, had yet to be discovered. Eddington managed to sidestep this problem by concentrating on the thermodynamics of radiative transport of energy in stellar interiors. So, Eddington predicted that dwarf stars remain in an essentially static position on the main sequence for most of their lives. In the 1930s and 1940s, with an understanding of hydrogen fusion, came a physically based theory of evolution to red giants, and white dwarfs. By this time, study of the Hertzsprung–Russell diagram did not drive such developments but merely allowed stellar evolution to be presented graphically.

Active Galactic Nucleus

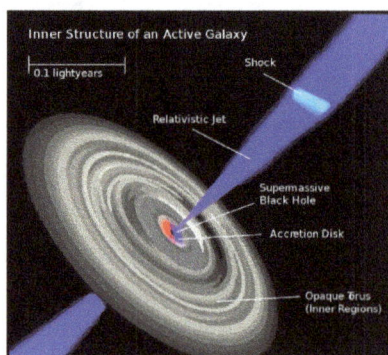

Inner Structure of an Active Galaxy

An active galactic nucleus (AGN) is a compact region at the center of a galaxy that has a much higher than normal luminosity over at least some portion – and possibly all – of the electromagnetic spectrum. Such excess emission has been observed in the radio, microwaves, infrared, optical, ultra-violet, X-ray and gamma ray wavebands. A galaxy hosting an AGN is called an active galaxy. The radiation from an AGN is believed to be a result of accretion of matter by a supermassive black hole at the center of its host galaxy. AGN are the most luminous persistent sources of electromagnetic radiation in the universe, and as such can be used as a means of discovering distant objects; their evolution as a function of cosmic time also puts constraints on models of the cosmos.

Models of the Active Nucleus

For a long time it has been argued that an AGN must be powered by accretion of mass onto massive black holes (10^6 to 10^{10} times the Solar mass). AGN are both compact and persistently extremely luminous. Accretion can potentially give very efficient conversion of potential and kinetic energy to radiation, and a massive black hole has a high Eddington luminosity, and as a result, it can provide the observed high persistent luminosity. Supermassive black holes are now believed to exist in the centres of most if not all massive galaxies since the mass of the black hole correlates well with the velocity dispersion of the galactic bulge (the M-sigma relation) or with bulge luminosity. Thus AGN-like characteristics are expected whenever a supply of material for accretion comes within the sphere of influence of the central black hole.

Accretion Disc

In the standard model of AGN, cold material close to a black hole forms an accretion disc. Dissipative processes in the accretion disc transport matter inwards and angular momentum outwards, while causing the accretion disc to heat up. The expected spectrum of an accretion disc peaks in the optical-ultraviolet waveband; in addition, a corona of hot material forms above the accretion disc and can inverse-Compton scatter photons up to X-ray energies. The radiation from the accretion disc excites cold atomic material close to the black hole and this in turn radiates at particular emission lines. A large fraction of the AGN's radiation may be obscured by interstellar gas and dust close to the accretion disc, but (in a steady-state situation) this will be re-radiated at some other waveband, most likely the infrared.

Relativistic Jets

Image taken by the Hubble Space Telescope of a 5000-light-year-long jet ejected from the active galaxy M87. The blue synchrotron radiation contrasts with the yellow starlight from the host galaxy.

Some accretion discs produce jets of twin, highly collimated, and fast outflows that emerge in opposite directions from close to the disc. The direction of the jet ejection is determined either by the angular momentum axis of the accretion disc or the spin axis of the black hole. The jet production mechanism and indeed the jet composition on very small scales are not understood at present due to the resolution of astronomical instruments being too low, and as a result, observations cannot provide enough evidence to support one of the various theoretical models of jet production over the many that exist. The jets have their most obvious observational effects in the radio waveband, where very-long-baseline interferometry can be used to study the synchrotron radiation they emit at resolutions of sub-parsec scales. However, they radiate in all wavebands from the radio through to the gamma-ray range via the synchrotron and the inverse-Compton scattering process, and so AGN jets are a second potential source of any observed continuum radiation.

Radiatively Inefficient AGN

There exists a class of 'radiatively inefficient' solutions to the equations that govern accretion. The most widely known of these is the Advection Dominated Accretion Flow (ADAF), but other theories exist. In this type of accretion, which is important for accretion rates well below the Eddington limit, the accreting matter does not form a thin disc and consequently does not efficiently radiate away the energy that it acquired as it moved close to the black hole. Radiatively inefficient accretion has been used to explain the lack of strong AGN-type radiation from massive black holes at the centres of elliptical galaxies in clusters, where otherwise we might expect high accretion rates and correspondingly high luminosities. Radiatively inefficient AGN would be expected to lack many of the characteristic features of standard AGN with an accretion disc.

Particle Acceleration

Observational Characteristics

There is no single observational signature of an AGN. The list below covers some of the historically important features that have allowed systems to be identified as AGN.

Nuclear optical continuum emission. This is visible whenever there is a direct view of the accretion disc. Jets can also contribute to this component of the AGN emission. The optical emission has a roughly power-law dependence on wavelength.

Nuclear infra-red emission. This is visible whenever the accretion disc and its environment are obscured by gas and dust close to the nucleus and then re-emitted ('reprocessing'). As it is thermal emission, it can be distinguished from any jet or disc-related emission.

Broad optical emission lines. These come from cold material close to the central black hole. The lines are broad because the emitting material is revolving around the black hole with high speeds causing a range of Doppler shifts of the emitted photons.

Narrow optical emission lines. These come from more distant cold material, and so are narrower than the broad lines.

Radio continuum emission. This is always due to a jet. It shows a spectrum characteristic of synchrotron radiation.

X-ray continuum emission. This can arise both from a jet and from the hot corona of the accretion disc via a scattering process: in both cases it shows a power-law spectrum. In some radio-quiet AGN there is an excess of soft X-ray emission in addition to the power-law component. The origin of the soft X-rays is not clear at present.

X-ray line emission. This is a result of illumination of cold heavy elements by the X-ray continuum that causes fluorescence of X-ray emission lines, the best-known of which is the iron feature around 6.4 keV. This line may be narrow or broad: relativistically broadened iron lines can be used to study the dynamics of the accretion disc very close to the nucleus and therefore the nature of the central black hole.

Types of Active Galaxy

It is convenient to divide AGN into two classes, conventionally called radio-quiet and radio-loud. Radio-loud objects have emission contributions from both the jet(s) and the lobes that the jets inflate. These emission contributions dominate the luminosity of the AGN at radio wavelengths and possibly at some or all other wavelengths. Radio-quiet objects are simpler since jet and any jet-related emission can be neglected at all wavelengths.

AGN terminology is often confusing, since the distinctions between different types of AGN sometimes reflect historical differences in how the objects were discovered or initially classified, rather than real physical differences.

Radio-quiet AGN

Low-ionization nuclear emission-line regions (LINERs). As the name suggests, these systems show only weak nuclear emission-line regions, and no other signatures of AGN emission. It is debatable whether all such systems are true AGN (powered by accretion on to a supermassive black hole). If they are, they constitute the lowest-luminosity class of radio-quiet AGN. Some may be radio-quiet analogues of the low-excitation radio galaxies.

Seyfert galaxies. Seyferts were the earliest distinct class of AGN to be identified. They show optical range nuclear continuum emission, narrow and occasionally broad emission lines, occasionally strong nuclear X-ray emission and sometimes a weak small-scale radio jet. Originally they were divided into two types known as Seyfert 1 and 2: Seyfert 1s show strong broad emission lines while Seyfert 2s do not, and Seyfert 1s are more likely to show strong low-energy X-ray emission. Various forms of elaboration on this scheme exist: for example, Seyfert 1s with relatively narrow broad lines are sometimes referred to as narrow-line Seyfert 1s. The host galaxies of Seyferts are usually spiral or irregular galaxies.

Radio-quiet quasars/QSOs. These are essentially more luminous versions of Seyfert 1s: the distinction is arbitrary and is usually expressed in terms of a limiting optical magnitude. Quasars were originally 'quasi-stellar' in optical images as they had optical luminosities that were greater than that of their host galaxy. They always show strong optical continuum emission, X-ray continuum emission, and broad and narrow optical emission lines. Some astronomers use the term QSO (Quasi-Stellar Object) for this class of AGN, reserving 'quasar' for radio-loud objects, while others talk about radio-quiet and radio-loud quasars. The host galaxies of quasars can be spirals, irregulars or ellipticals. There is a correlation between the quasar's luminosity and the mass of its host galaxy, in that the most luminous quasars inhabit the most massive galaxies (ellipticals).

'Quasar 2s'. By analogy with Seyfert 2s, these are objects with quasar-like luminosities but without strong optical nuclear continuum emission or broad line emission. They are scarce in surveys, though a number of possible candidate quasar 2s have been identified.

Radio-loud AGN

Radio-loud quasars behave exactly like radio-quiet quasars with the addition of emission from a jet. Thus they show strong optical continuum emission, broad and narrow emission lines, and strong X-ray emission, together with nuclear and often extended radio emission.

"Blazars" (BL Lac objects and OVV quasars) classes are distinguished by rapidly variable, polarized optical, radio and X-ray emission. BL Lac objects show no optical emission lines, broad or narrow, so that their redshifts can only be determined from features in the spectra of their host galaxies. The emission-line features may be intrinsically absent or simply swamped by the additional variable component. In the latter case, emission lines may become visible when the variable component is at a low level. OVV quasars behave more like standard radio-loud quasars with the addition of a rapidly variable component. In both classes of source, the variable emission is believed to originate in a relativistic jet oriented close to the line of sight. Relativistic effects amplify both the luminosity of the jet and the amplitude of variability.

Radio galaxies. These objects show nuclear and extended radio emission. Their other AGN properties are heterogeneous. They can broadly be divided into low-excitation and high-excitation classes. Low-excitation objects show no strong narrow or broad emission lines, and the emission lines they do have may be excited by a different mechanism. Their optical and X-ray nuclear emission is consistent with originating purely in a jet. They may be the best current candidates for AGN with radiatively inefficient accretion. By contrast, high-excitation objects (narrow-line radio galaxies) have emission-line spectra similar to those of Seyfert 2s. The small class of broad-line radio galaxies, which show relatively strong nuclear optical continuum emission probably includes some objects that are simply low-luminosity radio-loud quasars. The host galaxies of radio galaxies, whatever their emission-line type, are essentially always ellipticals.

Features of different types of galaxies										
Galaxy type	Active nuclei	Emission lines		X-rays	Excess of		Strong radio	Jets	Variable	Radio loud
		Narrow	Broad		UV	Far-IR				
Normal	no	weak	no	weak	no	no	no	no	no	no
LINER	unknown	weak	weak	weak	no	no	no	no	no	no
Seyfert I	yes	yes	yes	some	some	yes	few	no	yes	no
Seyfert II	yes	yes	no	some	some	yes	few	no	yes	no
Quasar	yes	yes	yes	some	yes	yes	some	some	yes	some
BL Lac	yes	no	no/faint	yes	yes	no	yes	yes	yes	yes
OVV	yes	no	stronger than BL Lac	yes	yes	no	yes	yes	yes	yes
Radio galaxy	yes	some	some	some	some	yes	yes	yes	yes	yes

Unification of AGN Species

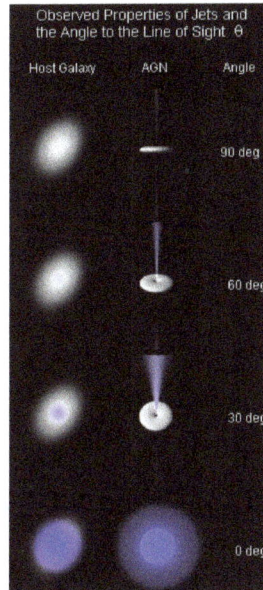

Unification by viewing angle. From bottom to top: *down the jet* - Blazar, *at an angle to the jet* - Quasar/Seyfert 1 Galaxy, *at 90 degrees from the jet* - Radio galaxy / Seyfert 2 Galaxy

Unified models propose that different observational classes of AGN are a single type of physical object observed under different conditions. The currently favoured unified models are 'orientation-based unified models' meaning that they propose that the apparent differences between different types of objects arise simply because of their different orientations to the observer. However, they are debated.

Radio-quiet Unification

At low luminosities, the objects to be unified are Seyfert galaxies. The unification models propose that in Seyfert 1s the observer has a direct view of the active nucleus. In Seyfert 2s the nucleus is observed through an obscuring structure which prevents a direct view of the optical continuum, broad-line region or (soft) X-ray emission. The key insight of orientation-dependent accretion models is that the two types of object can be the same if only certain angles to the line of sight are observed. The standard picture is of a torus of obscuring material surrounding the accretion disc. It must be large enough to obscure the broad-line region but not large enough to obscure the narrow-line region, which is seen in both classes of object. Seyfert 2s are seen through the torus. Outside the torus there is material that can scatter some of the nuclear emission into our line of sight, allowing us to see some optical and X-ray continuum and, in some cases, broad emission lines—which are strongly polarized, showing that they have been scattered and proving that some Seyfert 2s really do contain hidden Seyfert 1s. Infrared observations of the nuclei of Seyfert 2s also support this picture.

At higher luminosities, quasars take the place of Seyfert 1s, but, as already mentioned, the corresponding 'quasar 2s' are elusive at present. If they do not have the scattering component of Seyfert 2s they would be hard to detect except through their luminous narrow-line and hard X-ray emission.

Radio-loud Unification

Historically, work on radio-loud unification has concentrated on high-luminosity radio-loud quasars. These can be unified with narrow-line radio galaxies in a manner directly analogous to the Seyfert 1/2 unification (but without the complication of much in the way of a reflection component: narrow-line radio galaxies show no nuclear optical continuum or reflected X-ray component, although they do occasionally show polarized broad-line emission). The large-scale radio structures of these objects provide compelling evidence that the orientation-based unified models really are true. X-ray evidence, where available, supports the unified picture: radio galaxies show evidence of obscuration from a torus, while quasars do not, although care must be taken since radio-loud objects also have a soft unabsorbed jet-related component, and high resolution is necessary to separate out thermal emission from the sources' large-scale hot-gas environment. At very small angles to the line of sight, relativistic beaming dominates, and we see a blazar of some variety.

However, the population of radio galaxies is completely dominated by low-luminosity, low-excitation objects. These do not show strong nuclear emission lines — broad or narrow — they have optical continua which appear to be entirely jet-related, and their X-ray emission is also consistent with coming purely from a jet, with no heavily absorbed nuclear component in general. These objects cannot be unified with quasars, even though they include some high-luminosity objects when looking at radio emission, since the torus can never hide the narrow-line region to the required extent, and since infrared studies show that they have no hidden nuclear component: in fact there is no evidence for a torus in these objects at all. Most likely, they form a separate class in which only jet-related emission is important. At small angles to the line of sight, they will appear as BL Lac objects.

Criticism of the Radio-quiet Unification

In the recent literature on AGN, being subject to an intense debate, an increasing set of observations appear to be in conflict with some of the key predictions of the Unified Model, e.g. that each Seyfert 2 has an obscured Seyfert 1 nucleus (a hidden broad-line region).

Therefore, one cannot know whether the gas in all Seyfert 2 galaxies is ionized due to photoionization from a single, non-stellar continuum source in the center or due to shock-ionization from e.g. intense, nuclear starbursts. Spectropolarimetric studies reveal that only 50% of Seyfert 2s show a hidden broad-line region and thus split Seyfert 2 galaxies into two populations. The two classes of populations appear to differ by their luminosity, where the Seyfert 2s without a hidden broad-line region are generally less luminous. This suggests absence of broad-line region is connected to low Eddington ratio, and not to obscuration.

The covering factor of the torus might play an important role. Some torus models predict how Seyfert 1s and Seyfert 2s can obtain different covering factors from a luminosity- and accretion rate-dependence of the torus covering factor, something supported by studies in the x-ray of AGN. The models also suggest an accretion-rate dependence of the broad-line region and provide a natural evolution from more active engines in Seyfert 1s to more "dead" Seyfert 2s and can explain the observed break-down of the unified model at low luminosities and the evolution of the broad-line region.

While studies of single AGN show important deviations from the expectations of the unified model, results from statistical tests have been contradictory. The most important short-coming of statistical tests by direct comparisons of statistical samples of Seyfert 1s and Seyfert 2s is the introduction of selection biases due to anisotropic selection criteria.

Studying neighbour galaxies rather than the AGN themselves first suggested the numbers of neighbours were larger for Seyfert 2s than for Seyfert 1s, in contradiction with the Unified Model. Today, having overcome the previous limitations of small sample sizes and anisotropic selection, studies of neighbours of hundreds to thousands of AGN have shown that the neighbours of Seyfert 2s are intrinsically dustier and more star-forming than Seyfert 1s and a connection between AGN type, host galaxy morphology and collision history. Moreover, angular clustering studies of the two AGN types confirm that they reside in different environments and show that they reside within dark matter halos of different masses. The AGN environment studies are in line with evolution-based unification models where Seyfert 2s transform into Seyfert 1s during merger, supporting earlier models of merger-driven activation of Seyfert 1 nuclei.

While controversy about the soundness of each individual study still prevails, they all agree on that the simplest viewing-angle based models of AGN Unification are incomplete. While it still might be valid that an obscured Seyfert 1 can appear as a Seyfert 2, not all Seyfert 2s must host an obscured Seyfert 1. Understanding whether it is the same engine driving all Seyfert 2s, the connection to radio-loud AGN, the mechanisms of the variability of some AGN that vary between the two types at very short time scales, and the connection of the AGN type to small- and large-scale environment remain important issues to incorporate into any unified model of active galactic nuclei.

Cosmological uses and Evolution

For a long time, active galaxies held all the records for the highest-redshift objects known either in the optical or the radio spectrum, because of their high luminosity. They still have a role to play in studies of the early universe, but it is now recognised that an AGN gives a highly biased picture of the 'typical' high-redshift galaxy.

More interesting is the study of the evolution of the AGN population. Most luminous classes of AGN (radio-loud and radio-quiet) seem to have been much more numerous in the early universe. This suggests (1) that massive black holes formed early on and (2) that the conditions for the formation of luminous AGN were more common in the early universe, such as a much higher availability of cold gas near the centre of galaxies than at present. It also implies that many objects that were once luminous quasars are now much less luminous, or entirely quiescent. The evolution of the low-luminosity AGN population is much less well understood due to the difficulty of observing these objects at high redshifts.

References

- International Bureau of Weights and Measures (2006), The International System of Units (SI) (PDF) (8th ed.), pp. 112–13, ISBN 92-822-2213-6 .

- McCarthy, Dennis D.; Petit, Gérard, eds. (2004), "IERS Conventions (2003)", IERS Technical Note No. 32, Frankfurt: Bundesamts für Kartographie und Geodäsie, ISBN 3-89888-884-3

- E. Chaisson; S. McMillan (1995). Astronomy: a beginner's guide to the universe (2nd ed.). Upper Saddle River,

New Jersey: Prentice-Hall. ISBN 0-13-733916-X.

- Johansen, A.; Jacquet, E.; Cuzzi, J. N.; Morbidelli, A.; Gounelle, M. (2015). "New Paradigms For Asteroid Formation". In Michel, P.; DeMeo, F.; Bottke, W. Asteroids IV. Space Science Series. University of Arizona Press. p. 471. arXiv:1505.02941. Bibcode:2015arXiv150502941J. ISBN 978-0-8165-3213-1.

- D'Angelo, Gennaro; Durisen, Richard H.; Lissauer, Jack J. (December 2010). "Giant Planet Formation". In Seager, Sara. Exoplanets. University of Arizona Press. pp. 319–346. arXiv:1006.5486. Bibcode:2010cxop. book..319D. ISBN 978-0-8165-2945-2.

- Bennett, Jeffrey; Donahue, Megan; Schneider, Nicholas; Voit, Mark (2014). "Formation of the Solar System". The Cosmic Perspective (7th ed.). San Francisco: Pearson. pp. 136–169. ISBN 978-0-321-89384-0.

- Scott, Edward R. D. (2002). "Meteorite Evidence for the Accretion and Collisional Evolution of Asteroids" (PDF). In Bottke Jr., W. F.; Cellino, A.; Paolicchi, P.; Binzel, R. P. Asteroids III. University of Arizona Press. pp. 697–709. Bibcode:2002aste.conf..697S. ISBN 978-0-8165-2281-1.

- Shukolyukov, A.; Lugmair, G. W. (2002). "Chronology of Asteroid Accretion and Differentiation" (PDF). In Bottke Jr., W. F.; Cellino, A.; Paolicchi, P.; Binzel, R. P. Asteroids III. pp. 687–695. Bibcode:2002aste. conf..687S. ISBN 978-0-8165-2281-1.

- Levison, Harold F.; Donnes, Luke (2007). "Comet Populations and Cometary Dynamics". In McFadden, Lucy-Ann Adams; Weissman, Paul Robert; Johnson, Torrence V. Encyclopedia of the Solar System (2nd ed.). Amsterdam: Academic Press. pp. 575–588. ISBN 0-12-088589-1.

- Krishna Swamy, K. S. (May 1997). Physics of Comets. World Scientific Series in Astronomy and Astrophysics, Volume 2 (2nd ed.). World Scientific. p. 364. ISBN 981-02-2632-2.

- Filacchione, Gianrico; Capaccioni, Fabrizio; Taylor, Matt; Bauer, Markus (13 January 2016). "Exposed ice on Rosetta's comet confirmed as water" (Press release). European Space Agency. Retrieved 14 January 2016.

- "Evaporation and Accretion of Extrasolar Comets Following White Dwarf Kicks". Cornell University Department of Astronomy. 2014. Retrieved 22 January 2016.

4

Star: An Overview

A star is largely composed of gases and it produces heat, light, ultraviolent rays and other forms of radiations. The closest star to the earth is the Sun. This chapter discusses star formation, stellar evolution, supernova, and white dwarf. It elucidates the crucial theories and principles of stars.

Star

A star-forming region in the Large Magellanic Cloud.

False-color imagery of the Sun, a G-type main-sequence star, the closest to Earth

A star is a luminous sphere of plasma held together by its own gravity. The nearest star to Earth is the Sun. Many other stars are visible to the naked eye from Earth during the night, appearing as a multitude of fixed luminous points in the sky due to their immense distance from Earth. Historically, the most prominent stars were grouped into constellations and asterisms, the brightest of which gained proper names. Astronomers have assembled star catalogues that identify the known stars and provide standardized stellar designations. However, most of the stars in the Universe, including all stars outside our galaxy, the Milky Way, are invisible to the naked eye from Earth. Indeed, most are invisible from Earth even through the most powerful telescopes.

For at least a portion of its life, a star shines due to thermonuclear fusion of hydrogen into helium in its core, releasing energy that traverses the star's interior and then radiates into outer space. Almost all naturally occurring elements heavier than helium are created by stellar nucleosynthesis during the star's lifetime, and for some stars by supernova nucleosynthesis when it explodes. Near the end of its life, a star can also contain degenerate matter. Astronomers can determine the mass, age, metallicity (chemical composition), and many other properties of a star by observing its motion through space, its luminosity, and spectrum respectively. The total mass of a star is the main factor that determines its evolution and eventual fate. Other characteristics of a star, including

18
An Introduction to Astrophysics

diameter and temperature, change over its life, while the star's environment affects its rotation and movement. A plot of the temperature of many stars against their luminosities produces a plot known as a Hertzsprung–Russell diagram (H–R diagram). Plotting a particular star on that diagram allows the age and evolutionary state of that star to be determined.

A star's life begins with the gravitational collapse of a gaseous nebula of material composed primarily of hydrogen, along with helium and trace amounts of heavier elements. When the stellar core is sufficiently dense, hydrogen becomes steadily converted into helium through nuclear fusion, releasing energy in the process. The remainder of the star's interior carries energy away from the core through a combination of radiative and convective heat transfer processes. The star's internal pressure prevents it from collapsing further under its own gravity. A star with mass greater than 0.4 times the Sun's will expand to become a red giant when the hydrogen fuel in its core is exhausted. it In some cases, it will fuse heavier elements at the core or in shells around the core. As the star expands it throws a part of its mass, enriched with those heavier elements, into the interstellar environment, to be recycled later as new stars. Meanwhile, the core becomes a stellar remnant: a white dwarf, a neutron star, or if it is sufficiently massive a black hole.

Binary and multi-star systems consist of two or more stars that are gravitationally bound and generally move around each other in stable orbits. When two such stars have a relatively close orbit, their gravitational interaction can have a significant impact on their evolution. Stars can form part of a much larger gravitationally bound structure, such as a star cluster or a galaxy.

Observation History

People have seen patterns in the stars since ancient times. This 1690 depiction of the constellation of Leo, the lion, is by Johannes Hevelius.

Historically, stars have been important to civilizations throughout the world. They have been part of religious practices and used for celestial navigation and orientation. Many ancient astronomers believed that stars were permanently affixed to a heavenly sphere and that they were immutable. By convention, astronomers grouped stars into constellations and used them to track the motions of the planets and the inferred position of the Sun. The motion of the Sun against the background stars (and the horizon) was used to create calendars, which could be used to regulate agricultural practices. The Gregorian calendar, currently used nearly everywhere in the world, is a solar calendar based on the angle of the Earth's rotational axis relative to its local star, the Sun.

The oldest accurately dated star chart was the result of ancient Egyptian astronomy in 1534 BC. The earliest known star catalogues were compiled by the ancient Babylonian astronomers of Mesopotamia in the late 2nd millennium BC, during the Kassite Period (*ca.* 1531–1155 BC).

The constellation of Leo as it can be seen by the naked eye. Lines have been added.

The first star catalogue in Greek astronomy was created by Aristillus in approximately 300 BC, with the help of Timocharis. The star catalog of Hipparchus (2nd century BC) included 1020 stars, and was used to assemble Ptolemy's star catalogue. Hipparchus is known for the discovery of the first recorded *nova* (new star). Many of the constellations and star names in use today derive from Greek astronomy.

In spite of the apparent immutability of the heavens, Chinese astronomers were aware that new stars could appear. In 185 AD, they were the first to observe and write about a supernova, now known as the SN 185. The brightest stellar event in recorded history was the SN 1006 supernova, which was observed in 1006 and written about by the Egyptian astronomer Ali ibn Ridwan and several Chinese astronomers. The SN 1054 supernova, which gave birth to the Crab Nebula, was also observed by Chinese and Islamic astronomers.

Medieval Islamic astronomers gave Arabic names to many stars that are still used today and they invented numerous astronomical instruments that could compute the positions of the stars. They built the first large observatory research institutes, mainly for the purpose of producing *Zij* star catalogues. Among these, the *Book of Fixed Stars* (964) was written by the Persian astronomer Abd al-Rahman al-Sufi, who observed a number of stars, star clusters (including the Omicron Velorum and Brocchi's Clusters) and galaxies (including the Andromeda Galaxy). According to A. Zahoor, in the 11th century, the Persian polymath scholar Abu Rayhan Biruni described the Milky Way galaxy as a multitude of fragments having the properties of nebulous stars, and also gave the latitudes of various stars during a lunar eclipse in 1019.

According to Josep Puig, the Andalusian astronomer Ibn Bajjah proposed that the Milky Way was made up of many stars that almost touched one another and appeared to be a continuous image due to the effect of refraction from sublunary material, citing his observation of the conjunction of Jupiter and Mars on 500 AH (1106/1107 AD) as evidence. Early European astronomers such as Tycho Brahe identified new stars in the night sky (later termed *novae*), suggesting that the heavens were not immutable. In 1584 Giordano Bruno suggested that the stars were like the Sun, and may have other planets, possibly even Earth-like, in orbit around them, an idea that had been suggested earlier by the ancient Greek philosophers, Democritus and Epicurus, and by medieval Islamic cosmologists such as Fakhr al-Din al-Razi. By the following century, the idea of the stars being the same as the Sun was reaching a consensus among astronomers. To explain why these stars exerted no net gravitational pull on the Solar System, Isaac Newton suggested that the stars were equally distributed in every direction, an idea prompted by the theologian Richard Bentley.

The Italian astronomer Geminiano Montanari recorded observing variations in luminosity of the star Algol in 1667. Edmond Halley published the first measurements of the proper motion of a pair of nearby "fixed" stars, demonstrating that they had changed positions since the time of the ancient Greek astronomers Ptolemy and Hipparchus.

William Herschel was the first astronomer to attempt to determine the distribution of stars in the sky. During the 1780s he established a series of gauges in 600 directions and counted the stars observed along each line of sight. From this he deduced that the number of stars steadily increased toward one side of the sky, in the direction of the Milky Way core. His son John Herschel repeated this study in the southern hemisphere and found a corresponding increase in the same direction. In addition to his other accomplishments, William Herschel is also noted for his discovery that some stars do not merely lie along the same line of sight, but are also physical companions that form binary star systems.

The science of stellar spectroscopy was pioneered by Joseph von Fraunhofer and Angelo Secchi. By comparing the spectra of stars such as Sirius to the Sun, they found differences in the strength and number of their absorption lines—the dark lines in a stellar spectra caused by the atmosphere's absorption of specific frequencies. In 1865 Secchi began classifying stars into spectral types. However, the modern version of the stellar classification scheme was developed by Annie J. Cannon during the 1900s.

Alpha Centauri A and B over limb of Saturn

The first direct measurement of the distance to a star (61 Cygni at 11.4 light-years) was made in 1838 by Friedrich Bessel using the parallax technique. Parallax measurements demonstrated the vast separation of the stars in the heavens. Observation of double stars gained increasing importance during the 19th century. In 1834, Friedrich Bessel observed changes in the proper motion of the star Sirius and inferred a hidden companion. Edward Pickering discovered the first spectroscopic binary in 1899 when he observed the periodic splitting of the spectral lines of the star Mizar in a 104-day period. Detailed observations of many binary star systems were collected by astronomers such as William Struve and S. W. Burnham, allowing the masses of stars to be determined from computation of orbital elements. The first solution to the problem of deriving an orbit of binary stars from telescope observations was made by Felix Savary in 1827. The twentieth century saw increasingly rapid advances in the scientific study of stars. The photograph became a valuable astronomical tool. Karl Schwarzschild discovered that the color of a star and, hence, its temperature, could be determined by comparing the visual magnitude against the photographic magnitude. The development of the photoelectric photometer allowed precise measurements of magnitude at multiple wavelength intervals. In 1921 Albert A. Michelson made the first measurements of a stellar diameter using an interferometer on the Hooker telescope at Mount Wilson Observatory.

Important theoretical work on the physical structure of stars occurred during the first decades of the twentieth century. In 1913, the Hertzsprung-Russell diagram was developed, propelling the astrophysical study of stars. Successful models were developed to explain the interiors of stars and stellar evolution. Cecilia Payne-Gaposchkin first proposed that stars were made primarily of hydrogen and helium in her 1925 PhD thesis. The spectra of stars were further understood through advances in quantum physics. This allowed the chemical composition of the stellar atmosphere to be determined.

With the exception of supernovae, individual stars have primarily been observed in the Local Group, and especially in the visible part of the Milky Way (as demonstrated by the detailed star catalogues available for our galaxy). But some stars have been observed in the M100 galaxy of the Virgo Cluster, about 100 million light years from the Earth. In the Local Supercluster it is possible to see star clusters, and current telescopes could in principle observe faint individual stars in the Local Group. However, outside the Local Supercluster of galaxies, neither individual stars nor clusters of stars have been observed. The only exception is a faint image of a large star cluster containing hundreds of thousands of stars located at a distance of one billion light years—ten times further than the most distant star cluster previously observed.

Designations

This view contains blue stars known as "Blue stragglers", for their apparent location on the Hertzsprung–Russell diagram

The concept of a constellation was known to exist during the Babylonian period. Ancient sky watchers imagined that prominent arrangements of stars formed patterns, and they associated these with particular aspects of nature or their myths. Twelve of these formations lay along the band of the ecliptic and these became the basis of astrology. Many of the more prominent individual stars were also given names, particularly with Arabic or Latin designations.

As well as certain constellations and the Sun itself, individual stars have their own myths. To the represented various important deities, from which the names of the planets Mercury, Venus, Mars, Jupiter and Saturn were taken. (Uranus and Neptune were also Greek and Roman gods, but neither planet was known in Antiquity because of their low brightness. Their names were assigned by later astronomers.)

Circa 1600, the names of the constellations were used to name the stars in the corresponding regions of the sky. The German astronomer Johann Bayer created a series of star maps and applied Greek letters as designations to the stars in each constellation. Later a numbering system based on the star's right ascension was invented and added to John Flamsteed's star catalogue in his book *"Historia coelestis Britannica"* (the 1712 edition), whereby this numbering system came to be called *Flamsteed designation* or *Flamsteed numbering*.

The only internationally recognized authority for naming celestial bodies is the International Astronomical Union (IAU). A number of private companies sell names of stars, which the British Library calls an unregulated commercial enterprise. The IAU has disassociated itself from this commercial practice, and these names are neither recognized by the IAU nor used by them. One such star-naming company is the International Star Registry, which, during the 1980s, was accused of deceptive practice for making it appear that the assigned name was official. This now-discontinued ISR practice was informally labeled a scam and a fraud, and the New York City Department of Consumer Affairs issued a violation against ISR for engaging in a deceptive trade practice.

Units of Measurement

Although stellar parameters can be expressed in SI units or CGS units, it is often most convenient to express mass, luminosity, and radii in solar units, based on the characteristics of the Sun:

solar mass:	$M_\odot = 1.9891 \times 10^{30}$ kg
solar luminosity:	$L_\odot = 3.827 \times 10^{26}$ W
solar radius	$R_\odot = 6.960 \times 10^8$ m

Large lengths, such as the radius of a giant star or the semi-major axis of a binary star system, are often expressed in terms of the astronomical unit —approximately equal to the mean distance between the Earth and the Sun (150 million km or 93 million miles).

Formation and Evolution

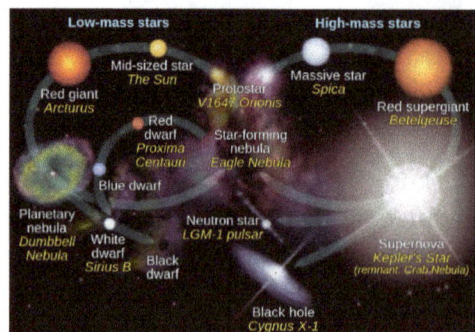

Stellar evolution of low-mass (left cycle) and high-mass (right cycle) stars, with examples in italics

Stars condense from regions of space of higher density, yet those regions are less dense than within a vacuum chamber. These regions - known as *molecular clouds* - consist mostly of hydrogen, with about 23 to 28 percent helium and a few percent heavier elements. One example of such a star-forming region is the Orion Nebula. Most stars form in groups of dozens to hundreds of thousands of stars. Massive stars in these groups may powerfully illuminate those clouds, ionizing the hydrogen, and creating H II regions. Such feedback effects, from star formation, may ultimately disrupt the cloud and prevent further star formation.

All stars spend the majority of their existence as *main sequence stars*, fueled primarily by the nuclear fusion of hydrogen into helium within their cores. However, stars of different masses have markedly different properties at various stages of their development. The ultimate fate of more massive stars differs from that of less massive stars, as do their luminosities and the impact they have on their environment. Accordingly, astronomers often group stars by their mass:

Very low mass stars, with masses below $0.5\,M_{\odot}$, are fully convective and distribute helium evenly throughout the whole star while on the main sequence. Therefore, they never undergo shell burning, never become red giants, which cease fusing and become helium white dwarfs and slowly cool after exhausting their hydrogen. However, as the lifetime of $0.5\,M_{\odot}$ stars is longer than the age of the universe, no such star has yet reached the white dwarf stage.

Low mass stars (including the Sun), with a mass between $0.5\,M_{\odot}$ and 1.8–$2.5\,M_{\odot}$ depending on composition, do become red giants as their core hydrogen is depleted and they begin to burn helium in core in a helium flash; they develop a degenerate carbon-oxygen core later on the asymptotic giant branch; they finally blow off their outer shell as a planetary nebula and leave behind their core in the form of a white dwarf.

Intermediate-mass stars, between 1.8–$2.5\,M_{\odot}$ and 5–$10\,M_{\odot}$, pass through evolutionary stages similar to low mass stars, but after a relatively short period on the RGB they ignite helium without a flash and spend an extended period in the red clump before forming a degenerate carbon-oxygen core.

Massive stars generally have a minimum mass of 7–$10\,M_{\odot}$ (possibly as low as 5–$6\,M_{\odot}$). After exhausting the hydrogen at the core these stars become supergiants and go on to fuse elements heavier than helium. They end their lives when their cores collapse and they explode as supernovae.

Star Formation

The formation of a star begins with gravitational instability within a molecular cloud, caused by regions of higher density - often triggered by compression of clouds by radiation from massive stars, expanding bubbles in the interstellar medium, the collision of different molecular clouds, or the collision of galaxies (as in a starburst galaxy). When a region reaches a sufficient density of matter to satisfy the criteria for Jeans instability, it begins to collapse under its own gravitational force.

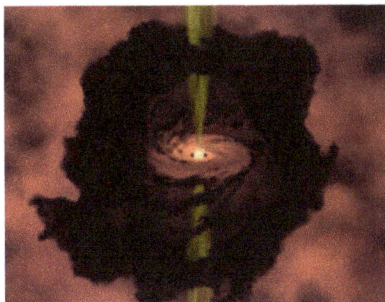

Artist's conception of the birth of a star within a dense molecular cloud.

As the cloud collapses, individual conglomerations of dense dust and gas form "Bok globules". As a globule collapses and the density increases, the gravitational energy converts into heat and the temperature rises. When the protostellar cloud has approximately reached the stable condition of hydrostatic equilibrium, a protostar forms at the core. These pre–main sequence stars are often surrounded by a protoplanetary disk and powered mainly by the conversion of gravitational energy. The period of gravitational contraction lasts about 10 to 15 million years.

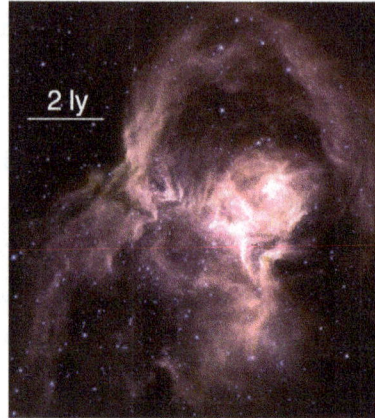

A cluster of approximately 500 young stars lies within the nearby W40 stellar nursery.

Early stars of less than $2\,M_\odot$ are called T Tauri stars, while those with greater mass are Herbig Ae/Be stars. These newly formed stars emit jets of gas along their axis of rotation, which may reduce the angular momentum of the collapsing star and result in small patches of nebulosity known as Herbig–Haro objects. These jets, in combination with radiation from nearby massive stars, may help to drive away the surrounding cloud from which the star was formed.

Early in their development, T Tauri stars follow the Hayashi track—they contract and decrease in luminosity while remaining at roughly the same temperature. Less massive T Tauri stars follow this track to the main sequence, while more massive stars turn onto the Henyey track.

Most stars are observed to be members of binary star systems, and the properties of those binaries are the result of the conditions in which they formed. A gas cloud must lose its angular momentum in order to collapse and form a star. The fragmentation of the cloud into multiple stars distributes some of that angular momentum. The primordial binaries transfer some angular momentum by gravitational interactions during close encounters with other stars in young stellar clusters. These interactions tend to split apart more widely separated (soft) binaries while causing hard binaries to become more tightly bound. This produces the separation of binaries into their two observed populations distributions.

Main Sequence

Stars spend about 90% of their existence fusing hydrogen into helium in high-temperature and high-pressure reactions near the core. Such stars are said to be on the main sequence, and are called dwarf stars. Starting at zero-age main sequence, the proportion of helium in a star's core will steadily increase, the rate of nuclear fusion at the core will slowly increase, as will the star's temperature and luminosity. The Sun, for example, is estimated to have increased in luminosity by about 40% since it reached the main sequence 4.6 billion (4.6×10^9) years ago.

Every star generates a stellar wind of particles that causes a continual outflow of gas into space. For most stars, the mass lost is negligible. The Sun loses 10^{-14} $M_\odot$ every year, or about 0.01% of its total mass over its entire lifespan. However, very massive stars can lose 10^{-7} to 10^{-5} $M_\odot$ each year, significantly affecting their evolution. Stars that begin with more than 50 $M_\odot$ can lose over half their total mass while on the main sequence.

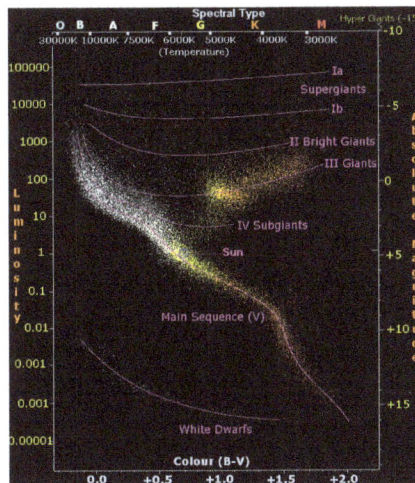

An example of a Hertzsprung–Russell diagram for a set of stars that includes the Sun (center).

The time a star spends on the main sequence depends primarily on the amount of fuel it has and the rate at which it fuses it. The Sun's is expected to live 10 billion (10^{10}) years. Massive stars consume their fuel very rapidly and are short-lived. Low mass stars consume their fuel very slowly. Stars less massive than 0.25 $M_\odot$, called red dwarfs, are able to fuse nearly all of their mass while stars of about 1 $M_\odot$ can only fuse about 10% of their mass. The combination of their slow fuel-consumption and relatively large usable fuel supply allows low mass stars to last about one trillion (10^{12}) years; the most extreme of 0.08 $M_\odot$) will last for about 12 trillion years. Red dwarfs become hotter and more luminous as they accumulate helium. When they eventually run out of hydrogen, they contract into a white dwarf and decline in temperature. However, since the lifespan of such stars is greater than the current age of the universe (13.8 billion years), no stars under about 0.85 $M_\odot$ are expected to have moved off the main sequence.

Besides mass, the elements heavier than helium can play a significant role in the evolution of stars. Astronomers label all elements heavier than helium "metals", and call the chemical concentration of these elements in a star, its metallicity. A star's metallicity can influence the time the star takes to burn its fuel, and controls the formation of its magnetic fields, which affects the strength of its stellar wind. Older, population II stars have substantially less metallicity than the younger, population I stars due to the composition of the molecular clouds from which they formed. Over time, such clouds become increasingly enriched in heavier elements as older stars die and shed portions of their atmospheres.

Post–Main Sequence

As stars of at least 0.4 $M_\odot$ exhaust their supply of hydrogen at their core, they start to fuse hydrogen in a shell outside the helium core. Their outer layers expand and cool greatly as they form a red giant. In about 5 billion years, when the Sun enters the helium burning phase, it will expand to a maximum radius of roughly 1 astronomical unit (150 million kilometres), 250 times its present size, and lose 30% of its current mass.

As the hydrogen shell burning produces more helium, the core increases in mass and temperature. In a red giant of up to 2.25 $M_\odot$, the mass of the helium core becomes degenerate prior to helium fusion. Finally, when the temperature increases sufficiently, helium fusion begins explosively in what is called a helium flash, and the star rapidly shrinks in radius, increases its surface temperature, and moves to the horizontal branch of the HR diagram. For more massive stars, helium core fusion starts before the core becomes degenerate, and the star spends some time in the red clump, slowly burning helium, before the outer convective envelope collapses and the star then moves to the horizontal branch.

After the star has fused the helium of its core, the carbon product fuses producing a hot core with an outer shell of fusing helium. The star then follows an evolutionary path called the asymptotic giant branch (AGB) that parallels the other described red giant phase, but with a higher luminosity. The more massive AGB stars may undergo a brief period of carbon fusion before the core becomes degenerate.

Massive Stars

During their helium-burning phase, stars of more than nine solar masses expand to form red supergiants. When this fuel is exhausted at the core, they continue to fuse elements heavier than helium.

The core contracts and the temperature and pressure rises enough to fuse carbon. This process continues, with the successive stages being fueled by neon, oxygen and silicon. Near the end of the star's life, fusion continues along a series of onion-layer shells within a massive star. Each shell fuses a different element, with the outermost shell fusing hydrogen; the next shell fusing helium, and so forth.

The final stage occurs when a massive star begins producing iron. Since iron nuclei are more tightly bound than any heavier nuclei, any fusion beyond iron does not produce a net release of energy. To a very limited degree such a process proceeds, but it consumes energy. Likewise, since they are more tightly bound than all lighter nuclei, such energy cannot be released by fission. In relatively old, very massive stars, a large core of inert iron will accumulate in the center of the star. The heavier elements in these stars can work their way to the surface, forming evolved objects known as Wolf-Rayet stars that have a dense stellar wind which sheds the outer atmosphere.

Collapse

As a star's core shrinks, the intensity of radiation from that surface increases, creating such radiation pressure on the outer shell of gas that it will push those layers away, forming a planetary nebula. If what remains after the outer atmosphere has been shed is less than 1.4 $M_\odot$, it shrinks to a relatively tiny object about the size of Earth, known as a white dwarf. White dwarfs lack the mass for further gravitational compression to take place. The electron-degenerate matter inside a white dwarf is no longer a plasma, even though stars are generally referred to as being spheres of plasma. Eventually, white dwarfs fade into black dwarfs over a very long period of time.

The Crab Nebula, remnants of a supernova that was first observed around 1050 AD

In larger stars, fusion continues until the iron core has grown so large (more than 1.4 $M_\odot$) that it can no longer support its own mass. This core will suddenly collapse as its electrons are driven into its protons, forming neutrons, neutrinos, and gamma rays in a burst of electron capture and inverse beta decay. The shockwave formed by this sudden collapse causes the rest of the star to explode in a supernova. Supernovae become so bright that they may briefly outshine the star's entire home galaxy. When they occur within the Milky Way, supernovae have historically been observed by naked-eye observers as "new stars" where none seemingly existed before.

A supernova explosion blows away the star's outer layers, leaving a remnant such as the Crab Nebula. The core is compressed into a neutron star, which sometimes manifests itself as a pulsar or X-ray burster. In the case of the largest stars, the remnant is a black hole greater than 4 $M_\odot$)s. In a neutron star the matter is in a state known as neutron-degenerate matter, with a more exotic form of degenerate matter, QCD matter, possibly present in the core. Within a black hole, the matter is in a state that is not currently understood.

The blown-off outer layers of dying stars include heavy elements, which may be recycled during the formation of new stars. These heavy elements allow the formation of rocky planets. The outflow from supernovae and the stellar wind of large stars play an important part in shaping the interstellar medium.

Binary Stars

The post–main-sequence evolution of binary stars may be significantly different from the evolution of single stars of the same mass. If stars in a binary system are sufficiently close, when one of the stars expands to become a red giant it may overflow its Roche lobe, the region around a star where material is gravitationally bound to that star, leading to transfer of material to the other. When the Roche lobe is violated, a variety of phenomena can result, including contact binaries, common-envelope binaries, cataclysmic variables, and type Ia supernovae.

Distribution

A white dwarf star in orbit around Sirius (artist's impression).

In addition to isolated stars, a multi-star system can consist of two or more gravitationally bound stars that orbit each other. The simplest and most common multi-star system is a binary star, but systems of three or more stars are also found. For reasons of orbital stability, such multi-star systems are often organized into hierarchical sets of binary stars. Larger groups called star clusters also exist. These range from loose stellar associations with only a few stars, up to enormous globular clusters with hundreds of thousands of stars. Such systems orbit our Milky Way galaxy.

It has been a long-held assumption that the majority of stars occur in gravitationally bound, multiple-star systems. This is particularly true for very massive O and B class stars, where 80% of the stars are believed to be part of multiple-star systems. The proportion of single star systems increases with decreasing star mass, so that only 25% of red dwarfs are known to have stellar companions. As 85% of all stars are red dwarfs, most stars in the Milky Way are likely single from birth.

Stars are not spread uniformly across the universe, but are normally grouped into galaxies along with interstellar gas and dust. A typical galaxy contains hundreds of billions of stars, and there are more than 100 billion (10^{11}) galaxies in the observable universe. In 2010, one estimate of the number of stars in the observable universe was 300 sextillion (3×10^{23}). While it is often believed that stars only exist within galaxies, intergalactic stars have been discovered.

The nearest star to the Earth, apart from the Sun, is Proxima Centauri, which is 39.9 trillion kilometres, or 4.2 light-years. Travelling at the orbital speed of the Space Shuttle (8 kilometres per second—almost 30,000 kilometres per hour), it would take about 150,000 years to arrive. This it typical of stellar separations in galactic discs. Stars can be much closer to each other in the centres of galaxies and in globular clusters, or much farther apart in galactic halos.

Due to the relatively vast distances between stars outside the galactic nucleus, collisions between stars are thought to be rare. In denser regions such as the core of globular clusters or the galactic center, collisions can be more common. Such collisions can produce what are known as blue stragglers. These abnormal stars have a higher surface temperature than the other main sequence stars with the same luminosity of the cluster to which it belongs.

Characteristics

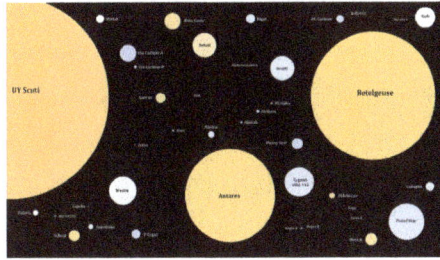

Some of the well-known stars with their apparent colors and relative sizes.

Almost everything about a star is determined by its initial mass, including such characteristics as luminosity, size, evolution, lifespan, and its eventual fate.

Age

Most stars are between 1 billion and 10 billion years old. Some stars may even be close to 13.8 billion years old—the observed age of the universe. The oldest star yet discovered, HD 140283, nicknamed Methuselah star, is an estimated 14.46 ± 0.8 billion years old. (Due to the uncertainty in the value, this age for the star does not conflict with the age of the Universe, determined by the Planck satellite as 13.799 ± 0.021).

The more massive the star, the shorter its lifespan, primarily because massive stars have greater pressure on their cores, causing them to burn hydrogen more rapidly. The most massive stars last an average of a few million years, while stars of minimum mass (red dwarfs) burn their fuel very slowly and can last tens to hundreds of billions of years.

Chemical Composition

When stars form in the present Milky Way galaxy they are composed of about 71% hydrogen and 27% helium, as measured by mass, with a small fraction of heavier elements. Typically the portion of heavy elements is measured in terms of the iron content of the stellar atmosphere, as iron is a common element and its absorption lines are relatively easy to measure. The portion of heavier elements may be an indicator of the likelihood that the star has a planetary system.

The star with the lowest iron content ever measured is the dwarf HE1327-2326, with only 1/200,000th the iron content of the Sun. By contrast, the super-metal-rich star μ Leonis has nearly double the abundance of iron as the Sun, while the planet-bearing star 14 Herculis has nearly triple the iron. There also exist chemically peculiar stars that show unusual abundances of certain elements in their spectrum; especially chromium and rare earth elements. Stars with cooler outer atmospheres, including the Sun, can form various diatomic and polyatomic molecules.

Diameter

Due to their great distance from the Earth, all stars except the Sun appear to the unaided eye as shining points in the night sky that twinkle because of the effect of the Earth's atmosphere. The

Sun is also a star, but it is close enough to the Earth to appear as a disk instead, and to provide daylight. Other than the Sun, the star with the largest apparent size is R Doradus, with an angular diameter of only 0.057 arcseconds.

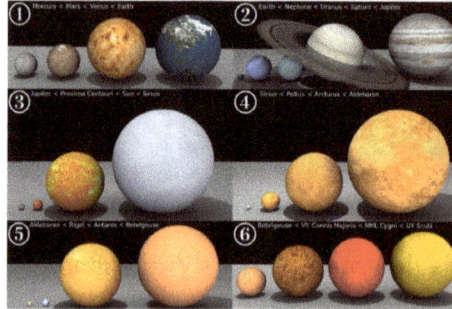

Stars vary widely in size. In each image in the sequence, the right-most object appears as the left-most object in the next panel. The Earth appears at right in panel 1 and the Sun is second from the right in panel 3. The rightmost star at panel 6 is UY Scuti, the largest known star.

The disks of most stars are much too small in angular size to be observed with current ground-based optical telescopes, and so interferometer telescopes are required to produce images of these objects. Another technique for measuring the angular size of stars is through occultation. By precisely measuring the drop in brightness of a star as it is occulted by the Moon (or the rise in brightness when it reappears), the star's angular diameter can be computed.

Stars range in size from neutron stars, which vary anywhere from 20 to 40 km (25 mi) in diameter, to supergiants like Betelgeuse in the Orion constellation, which has a diameter approximately 1,070 times that of the Sun—about 1,490,171,880 km (925,949,878 mi). Betelgeuse, however, has a much lower density than the Sun.

Kinematics

The Pleiades, an open cluster of stars in the constellation of Taurus. These stars share a common motion through space.

The motion of a star relative to the Sun can provide useful information about the origin and age of a star, as well as the structure and evolution of the surrounding galaxy. The components of motion of a star consist of the radial velocity toward or away from the Sun, and the traverse angular movement, which is called its proper motion.

Radial velocity is measured by the doppler shift of the star's spectral lines, and is given in units of km/s. The proper motion of a star, its parallax, is determined by precise astrometric measurements in units of milli-arc seconds (mas) per year. With knowledge of the star's parallax and its distance, the proper motion velocity can be calculated. Together with the radial velocity, the total velocity can be calculated. Stars with high rates of proper motion are likely to be relatively close to the Sun, making them good candidates for parallax measurements.

When both rates of movement are known, the space velocity of the star relative to the Sun or the galaxy can be computed. Among nearby stars, it has been found that younger population I stars have generally lower velocities than older, population II stars. The latter have elliptical orbits that are inclined to the plane of the galaxy. A comparison of the kinematics of nearby stars has allowed astronomers to trace their origin to common points in giant molecular clouds, and are referred to as stellar associations.

Magnetic Field

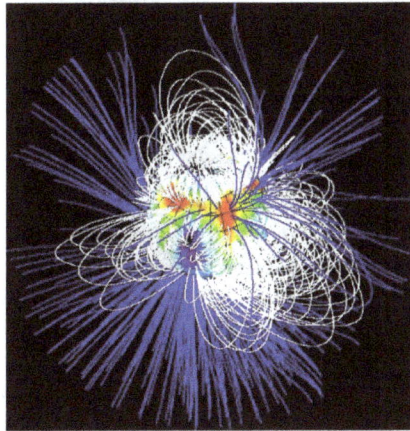

Surface magnetic field of SU Aur (a young star of T Tauri type), reconstructed by means of Zeeman-Doppler imaging

The magnetic field of a star is generated within regions of the interior where convective circulation occurs. This movement of conductive plasma functions like a dynamo, wherein the movement of electrical charges induce magnetic fields, as does a mechanical dynamo. Those magnetic fields have a great range that extend throughout and beyond the star. The strength of the magnetic field varies with the mass and composition of the star, and the amount of magnetic surface activity depends upon the star's rate of rotation. This surface activity produces starspots, which are regions of strong magnetic fields and lower than normal surface temperatures. Coronal loops are arching magnetic field flux lines that rise from a star's surface into the star's outer atmosphere, its corona. The coronal loops can be seen due to the plasma they conduct along their length. Stellar flares are bursts of high-energy particles that are emitted due to the same magnetic activity.

Young, rapidly rotating stars tend to have high levels of surface activity because of their magnetic field. The magnetic field can act upon a star's stellar wind, functioning as a brake to gradually slow the rate of rotation with time. Thus, older stars such as the Sun have a much slower rate of rotation and a lower level of surface activity. The activity levels of slowly rotating stars tend to vary in a cyclical manner and can shut down altogether for periods of time. During the Maunder minimum, for example, the Sun underwent a 70-year period with almost no sunspot activity.

Mass

One of the most massive stars known is Eta Carinae, which, with 100–150 times as much mass as the Sun, will have a lifespan of only several million years. Studies of the most massive open clusters suggests 150 $M_\odot$ as an upper limit for stars in the current era of the universe. This represents an empirical value for the theoretical limit on the mass of forming stars due to increasing radiation pressure on the accreting gas cloud. Several stars in the R136 cluster in the Large Magellanic Cloud have been measured with larger masses, but it has been determined that they could have been created through the collision and merger of massive stars in close binary systems, sidestepping the 150 $M_\odot$ limit on massive star formation.

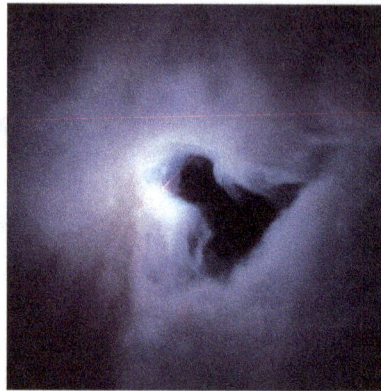

The reflection nebula NGC 1999 is brilliantly illuminated by V380 Orionis (center), a variable star with about 3.5 times the mass of the Sun. The black patch of sky is a vast hole of empty space and not a dark nebula as previously thought.

The first stars to form after the Big Bang may have been larger, up to 300 $M_\odot$, due to the complete absence of elements heavier than lithium in their composition. This generation of supermassive population III stars is likely to have existed in the very early universe (i.e., they are observed to have a high redshift), and may have started the production of chemical elements heavier than hydrogen that are needed for the later formation of planets and life. In June 2015, astronomers reported evidence for Population III stars in the Cosmos Redshift 7 galaxy at $z = 6.60$.

With a mass only 80 times that of Jupiter (M_J), 2MASS J0523-1403 is the smallest known star undergoing nuclear fusion in its core. For stars with metallicity similar to the Sun, the theoretical minimum mass the star can have and still undergo fusion at the core, is estimated to be about 75 M_J. When the metallicity is very low, however, the minimum star size seems to be about 8.3% of the solar mass, or about 87 M_J. Smaller bodies called brown dwarfs, occupy a poorly defined grey area between stars and gas giants.

The combination of the radius and the mass of a star determines its surface gravity. Giant stars have a much lower surface gravity than do main sequence stars, while the opposite is the case for degenerate, compact stars such as white dwarfs. The surface gravity can influence the appearance of a star's spectrum, with higher gravity causing a broadening of the absorption lines.

Rotation

The rotation rate of stars can be determined through spectroscopic measurement, or more exactly determined by tracking their starspots. Young stars can have a rotation greater than

100 km/s at the equator. The B-class star Achernar, for example, has an equatorial velocity of about 225 km/s or greater, causing its equator to be slung outward and giving it an equatorial diameter that is more than 50% greater than between the poles. This rate of rotation is just below the critical velocity of 300 km/s at which speed the star would break apart. By contrast, the Sun rotates once every 25 – 35 days, with an equatorial velocity of 1.994 km/s. A main sequence star's magnetic field and the stellar wind serve to slow its rotation by a significant amount as it evolves on the main sequence.

Degenerate stars have contracted into a compact mass, resulting in a rapid rate of rotation. However they have relatively low rates of rotation compared to what would be expected by conservation of angular momentum—the tendency of a rotating body to compensate for a contraction in size by increasing its rate of spin. A large portion of the star's angular momentum is dissipated as a result of mass loss through the stellar wind. In spite of this, the rate of rotation for a pulsar can be very rapid. The pulsar at the heart of the Crab nebula, for example, rotates 30 times per second. The rotation rate of the pulsar will gradually slow due to the emission of radiation.

Temperature

The surface temperature of a main sequence star is determined by the rate of energy production of its core and by its radius, and is often estimated from the star's color index. The temperature is normally given in terms of an effective temperature, which is the temperature of an idealized black body that radiates its energy at the same luminosity per surface area as the star. Note that the effective temperature is only a representative of the surface, as the temperature increases toward the core. The temperature in the core region of a star is several million kelvins.

The stellar temperature will determine the rate of ionization of various elements, resulting in characteristic absorption lines in the spectrum. The surface temperature of a star, along with its visual absolute magnitude and absorption features, is used to classify a star.

Massive main sequence stars can have surface temperatures of 50,000 K. Smaller stars such as the Sun have surface temperatures of a few thousand K. Red giants have relatively low surface temperatures of about 3,600 K; but they also have a high luminosity due to their large exterior surface area.

Radiation

The energy produced by stars, a product of nuclear fusion, radiates to space as both electromagnetic radiation and particle radiation. The particle radiation emitted by a star is manifested as the stellar wind, which streams from the outer layers as electrically charged protons and alpha and beta particles. Although almost massless, there also exists a steady stream of neutrinos emanating from the star's core.

The production of energy at the core is the reason stars shine so brightly: every time two or more atomic nuclei fuse together to form a single atomic nucleus of a new heavier element, gamma ray photons are released from the nuclear fusion product. This energy is converted to other forms of electromagnetic energy of lower frequency, such as visible light, by the time it reaches the star's outer layers.

The color of a star, as determined by the most intense frequency of the visible light, depends on the temperature of the star's outer layers, including its photosphere. Besides visible light, stars also emit forms of electromagnetic radiation that are invisible to the human eye. In fact, stellar electromagnetic radiation spans the entire electromagnetic spectrum, from the longest wavelengths of radio waves through infrared, visible light, ultraviolet, to the shortest of X-rays, and gamma rays. From the standpoint of total energy emitted by a star, not all components of stellar electromagnetic radiation are significant, but all frequencies provide insight into the star's physics.

Using the stellar spectrum, astronomers can also determine the surface temperature, surface gravity, metallicity and rotational velocity of a star. If the distance of the star is found, such as by measuring the parallax, then the luminosity of the star can be derived. The mass, radius, surface gravity, and rotation period can then be estimated based on stellar models. (Mass can be calculated for stars in binary systems by measuring their orbital velocities and distances. Gravitational microlensing has been used to measure the mass of a single star.) With these parameters, astronomers can also estimate the age of the star.

Luminosity

The luminosity of a star is the amount of light and other forms of radiant energy it radiates per unit of time. It has units of power. The luminosity of a star is determined by its radius and surface temperature. Many stars do not radiate uniformly across their entire surface. The rapidly rotating star Vega, for example, has a higher energy flux (power per unit area) at its poles than along its equator.

Patches of the star's surface with a lower temperature and luminosity than average are known as starspots. Small, *dwarf* stars such as our Sun generally have essentially featureless disks with only small starspots. *Giant* stars have much larger, more obvious starspots, and they also exhibit strong stellar limb darkening. That is, the brightness decreases towards the edge of the stellar disk. Red dwarf flare stars such as UV Ceti may also possess prominent starspot features.

Magnitude

The apparent brightness of a star is expressed in terms of its apparent magnitude. It is a function of the star's luminosity, its distance from Earth, and the altering of the star's light as it passes through Earth's atmosphere. Intrinsic or absolute magnitude is directly related to a star's luminosity, and is what the apparent magnitude a star would be if the distance between the Earth and the star were 10 parsecs (32.6 light-years).

Number of stars brighter than magnitude	
Apparent magnitude	Number of stars
0	4
1	15
2	48
3	171
4	513

5	1,602
6	4,800
7	14,000

Both the apparent and absolute magnitude scales are logarithmic units: one whole number difference in magnitude is equal to a brightness variation of about 2.5 times (the 5th root of 100 or approximately 2.512). This means that a first magnitude star (+1.00) is about 2.5 times brighter than a second magnitude (+2.00) star, and about 100 times brighter than a sixth magnitude star (+6.00). The faintest stars visible to the naked eye under good seeing conditions are about magnitude +6.

On both apparent and absolute magnitude scales, the smaller the magnitude number, the brighter the star; the larger the magnitude number, the fainter the star. The brightest stars, on either scale, have negative magnitude numbers. The variation in brightness (ΔL) between two stars is calculated by subtracting the magnitude number of the brighter star (m_b) from the magnitude number of the fainter star (m_f), then using the difference as an exponent for the base number 2.512; that is to say:

$$\Delta m = m_f - m_b$$

$$2.512^{\Delta m} = \Delta L$$

Relative to both luminosity and distance from Earth, a star's absolute magnitude (M) and apparent magnitude (m) are not equivalent; for example, the bright star Sirius has an apparent magnitude of −1.44, but it has an absolute magnitude of +1.41.

The Sun has an apparent magnitude of −26.7, but its absolute magnitude is only +4.83. Sirius, the brightest star in the night sky as seen from Earth, is approximately 23 times more luminous than the Sun, while Canopus, the second brightest star in the night sky with an absolute magnitude of −5.53, is approximately 14,000 times more luminous than the Sun. Despite Canopus being vastly more luminous than Sirius, however, Sirius appears brighter than Canopus. This is because Sirius is merely 8.6 light-years from the Earth, while Canopus is much farther away at a distance of 310 light-years.

As of 2006, the star with the highest known absolute magnitude is LBV 1806-20, with a magnitude of −14.2. This star is at least 5,000,000 times more luminous than the Sun. The least luminous stars that are currently known are located in the NGC 6397 cluster. The faintest red dwarfs in the cluster were magnitude 26, while a 28th magnitude white dwarf was also discovered. These faint stars are so dim that their light is as bright as a birthday candle on the Moon when viewed from the Earth.

Classification

Surface temperature ranges for different stellar classes		
Class	Temperature	Sample star
O	33,000 K or more	Zeta Ophiuchi
B	10,500–30,000 K	Rigel
A	7,500–10,000 K	Altair

F	6,000–7,200 K	Procyon A
G	5,500–6,000 K	Sun
K	4,000–5,250 K	Epsilon Indi
M	2,600–3,850 K	Proxima Centauri

The current stellar classification system originated in the early 20th century, when stars were classified from *A* to *Q* based on the strength of the hydrogen line. It thought that the hydrogen line strength was a simple linear function of temperature. Rather, it was more complicated; it strengthened with increasing temperature, it peaked near 9000 K, and then declined at greater temperatures. When the classifications were reordered by temperature, it more closely resembled the modern scheme.

Stars are given a single-letter classification according to their spectra, ranging from type *O*, which are very hot, to *M*, which are so cool that molecules may form in their atmospheres. The main classifications in order of decreasing surface temperature are: *O, B, A, F, G, K*, and *M*. A variety of rare spectral types are given special classifications. The most common of these are types *L* and *T*, which classify the coldest low-mass stars and brown dwarfs. Each letter has 10 sub-divisions, numbered from 0 to 9, in order of decreasing temperature. However, this system breaks down at extreme high temperatures as classes *O0* and *O1* may not exist.

In addition, stars may be classified by the luminosity effects found in their spectral lines, which correspond to their spatial size and is determined by their surface gravity. These range from *o* (hypergiants) through *III* (giants) to *V* (main sequence dwarfs); some authors add *VII* (white dwarfs). Most stars belong to the main sequence, which consists of ordinary hydrogen-burning stars. These fall along a narrow, diagonal band when graphed according to their absolute magnitude and spectral type. The Sun is a main sequence *G2V* yellow dwarf of intermediate temperature and ordinary size.

Additional nomenclature, in the form of lower-case letters added to the end of the spectral type to indicate peculiar features of the spectrum. For example, an "*e*" can indicate the presence of emission lines; "*m*" represents unusually strong levels of metals, and "*var*" can mean variations in the spectral type.

White dwarf stars have their own class that begins with the letter *D*. This is further sub-divided into the classes *DA, DB, DC, DO, DZ*, and *DQ*, depending on the types of prominent lines found in the spectrum. This is followed by a numerical value that indicates the temperature.

Variable Stars

Variable stars have periodic or random changes in luminosity because of intrinsic or extrinsic properties. Of the intrinsically variable stars, the primary types can be subdivided into three principal groups.

During their stellar evolution, some stars pass through phases where they can become pulsating variables. Pulsating variable stars vary in radius and luminosity over time, expanding and contracting with periods ranging from minutes to years, depending on the size of the star. This category includes Cepheid and Cepheid-like stars, and long-period variables such as Mira.

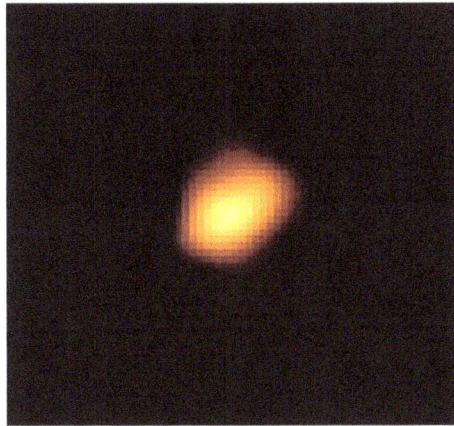

The asymmetrical appearance of Mira, an oscillating variable star.

Eruptive variables are stars that experience sudden increases in luminosity because of flares or mass ejection events. This group includes protostars, Wolf-Rayet stars, and flare stars, as well as giant and supergiant stars.

Cataclysmic or explosive variable stars are those that undergo a dramatic change in their properties. This group includes novae and supernovae. A binary star system that includes a nearby white dwarf can produce certain types of these spectacular stellar explosions, including the nova and a Type 1a supernova. The explosion is created when the white dwarf accretes hydrogen from the companion star, building up mass until the hydrogen undergoes fusion. Some novae are also recurrent, having periodic outbursts of moderate amplitude.

Stars can also vary in luminosity because of extrinsic factors, such as eclipsing binaries, as well as rotating stars that produce extreme starspots. A notable example of an eclipsing binary is Algol, which regularly varies in magnitude from 2.3 to 3.5 over a period of 2.87 days.

Structure

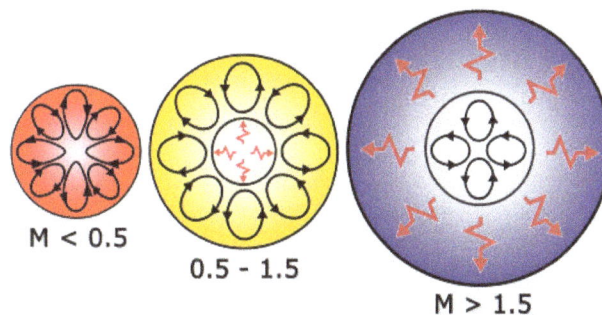

Internal structures of main sequence stars, convection zones with arrowed cycles and radiative zones with red flashes. To the left a **low-mass** red dwarf, in the center a **mid-sized** yellow dwarf, and, at the right, a **massive** blue-white main sequence star.

The interior of a stable star is in a state of hydrostatic equilibrium: the forces on any small volume almost exactly counterbalance each other. The balanced forces are inward gravitational force and an outward force due to the pressure gradient within the star. The pressure gradient is established by the temperature gradient of the plasma; the outer part of the star is cooler than the core. The temperature at the core of a main sequence or giant star is at least on the order of 10^7 K. The resulting temperature and pressure at the hydrogen-burning core of a main sequence star are sufficient for nuclear fusion to occur and for sufficient energy to be produced to prevent further collapse of the star.

As atomic nuclei are fused in the core, they emit energy in the form of gamma rays. These photons interact with the surrounding plasma, adding to the thermal energy at the core. Stars on the main sequence convert hydrogen into helium, creating a slowly but steadily increasing proportion of helium in the core. Eventually the helium content becomes predominant, and energy production ceases at the core. Instead, for stars of more than 0.4 $M_\odot$, fusion occurs in a slowly expanding shell around the degenerate helium core.

In addition to hydrostatic equilibrium, the interior of a stable star will also maintain an energy balance of thermal equilibrium. There is a radial temperature gradient throughout the interior that results in a flux of energy flowing toward the exterior. The outgoing flux of energy leaving any layer within the star will exactly match the incoming flux from below.

The radiation zone is the region of the stellar interior where the flux of energy outward is dependent on radiative heat transfer, since convective heat transfer is inefficient in that zone. In this region the plasma will not be perturbed, and any mass motions will die out. If this is not the case, however, then the plasma becomes unstable and convection will occur, forming a convection zone. This can occur, for example, in regions where very high energy fluxes occur, such as near the core or in areas with high opacity (making radiatative heat transfer inefficient) as in the outer envelope.

The occurrence of convection in the outer envelope of a main sequence star depends on the star's mass. Stars with several times the mass of the Sun have a convection zone deep within the interior and a radiative zone in the outer layers. Smaller stars such as the Sun are just the opposite, with the convective zone located in the outer layers. Red dwarf stars with less than 0.4 $M_\odot$ are convective throughout, which prevents the accumulation of a helium core. For most stars the convective zones will also vary over time as the star ages and the constitution of the interior is modified.

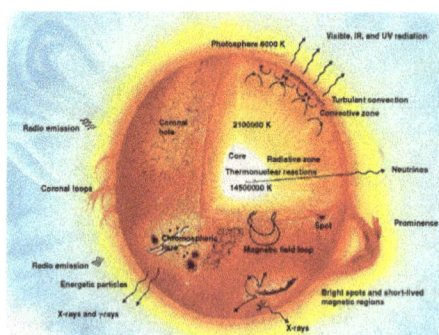

This diagram shows a cross-section of the Sun.

The photosphere is that portion of a star that is visible to an observer. This is the layer at which the plasma of the star becomes transparent to photons of light. From here, the energy generated at the core becomes free to propagate into space. It is within the photosphere that sun spots, regions of lower than average temperature, appear.

Above the level of the photosphere is the stellar atmosphere. In a main sequence star such as the Sun, the lowest level of the atmosphere, just above the photosphere, is the thin chromosphere region, where spicules appear and stellar flares begin. Above this is the transition region, where the temperature rapidly increases within a distance of only 100 km (62 mi). Beyond this is the corona, a volume of super-heated plasma that can extend outward to several million kilometres. The existence of a corona appears to be dependent on a convective zone in the outer layers of the star. Despite its high temperature, and the corona emits very little light, due to its low gas density. The corona region of the Sun is normally only visible during a solar eclipse.

From the corona, a stellar wind of plasma particles expands outward from the star, until it interacts with the interstellar medium. For the Sun, the influence of its solar wind extends throughout a bubble-shaped region called the heliosphere.

Nuclear Fusion Reaction Pathways

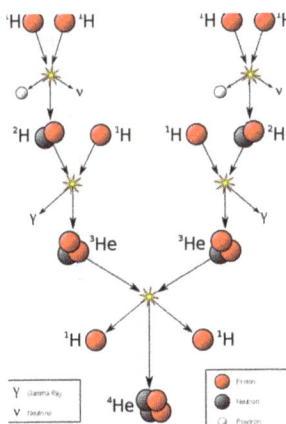

Overview of the proton-proton chain

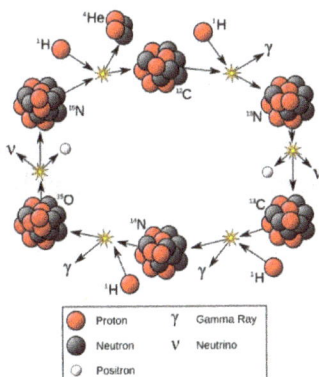

The carbon-nitrogen-oxygen cycle

A variety of nuclear fusion reactions take place in the cores of stars, that depend upon their mass and composition. When nuclei fuse, the mass of the fused product is less than the mass of the original parts. This lost mass is converted to electromagnetic energy, according to the mass-energy equivalence relationship $E = mc^2$.

The hydrogen fusion process is temperature-sensitive, so a moderate increase in the core temperature will result in a significant increase in the fusion rate. As a result, the core temperature of main sequence stars only varies from 4 million kelvin for a small M-class star to 40 million kelvin for a massive O-class star.

In the Sun, with a 10-million-kelvin core, hydrogen fuses to form helium in the proton-proton chain reaction:

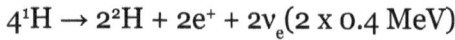

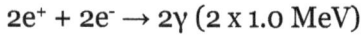

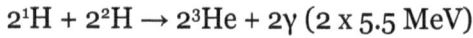

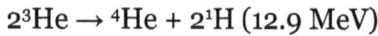

$$4^1H \rightarrow 2\,^2H + 2e^+ + 2\nu_e\,(2 \times 0.4\ MeV)$$

$$2e^+ + 2e^- \rightarrow 2\gamma\ (2 \times 1.0\ MeV)$$

$$2^1H + 2^2H \rightarrow 2^3He + 2\gamma\ (2 \times 5.5\ MeV)$$

$$2^3He \rightarrow {}^4He + 2^1H\ (12.9\ MeV)$$

These reactions result in the overall reaction:

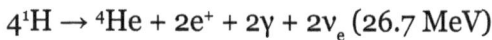

$$4^1H \rightarrow {}^4He + 2e^+ + 2\gamma + 2\nu_e\ (26.7\ MeV)$$

where e^+ is a positron, γ is a gamma ray photon, ν_e is a neutrino, and H and He are isotopes of hydrogen and helium, respectively. The energy released by this reaction is in millions of electron volts, which is actually only a tiny amount of energy. However enormous numbers of these reactions occur constantly, producing all the energy necessary to sustain the star's radiation output. In comparison, the combustion of two hydrogen gas molecules with one oxygen gas molecule releases only 5.7 eV.

Minimum stellar mass required for fusion	
Element	Solar masses
Hydrogen	0.01
Helium	0.4
Carbon	5
Neon	8

In more massive stars, helium is produced in a cycle of reactions catalyzed by carbon called the carbon-nitrogen-oxygen cycle.

In evolved stars with cores at 100 million kelvin and masses between 0.5 and 10 $M_\odot$, helium can be transformed into carbon in the triple-alpha process that uses the intermediate element beryllium:

$$^4He + {}^4He + 92\ keV \rightarrow {}^{8*}Be$$

$$^4He + {}^{8*}Be + 67\ keV \rightarrow {}^{12*}C$$

$$^{12*}C \rightarrow {}^{12}C + \gamma + 7.4\ MeV$$

For an overall reaction of:

$$3^4He \rightarrow {}^{12}C + \gamma + 7.2\ MeV$$

In massive stars, heavier elements can also be burned in a contracting core through the neon burning process and oxygen burning process. The final stage in the stellar nucleosynthesis process is the silicon burning process that results in the production of the stable isotope iron-56, an endothermic process that consumes energy, and so further energy can only be produced through gravitational collapse.

The example below shows the amount of time required for a star of 20 $M_\odot$ to consume all of its nuclear fuel. As an O-class main sequence star, it would be 8 times the solar radius and 62,000 times the Sun's luminosity.

Fuel material	Temperature (million kelvins)	Density (kg/cm³)	Burn duration (τ in years)
H	37	0.0045	8.1 million
He	188	0.97	1.2 million
C	870	170	976
Ne	1,570	3,100	0.6
O	1,980	5,550	1.25
S/Si	3,340	33,400	0.0315

Star Formation

Star formation is the process by which dense regions within molecular clouds in interstellar space, sometimes referred to as "stellar nurseries" or "star-forming regions", fuse to form stars. As a branch of astronomy, star formation includes the study of the interstellar medium (ISM) and giant molecular clouds (GMC) as precursors to the star formation process, and the study of protostars and young stellar objects as its immediate products. It is closely related to planet formation, another branch of astronomy. Star formation theory, as well as accounting for the formation of a single star, must also account for the statistics of binary stars and the initial mass function.

In June 2015, astronomers reported evidence for Population III stars in the Cosmos Redshift 7 galaxy at $z = 6.60$. Such stars are likely to have existed in the very early universe (i.e., at high redshift), and may have started the production of chemical elements heavier than hydrogen that are needed for the later formation of planets and life as we know it.

Stellar Nurseries

Hubble telescope image known as *Pillars of Creation,* where stars are forming in the Eagle Nebula.

Interstellar clouds

A spiral galaxy like the Milky Way contains stars, stellar remnants, and a diffuse interstellar medium (ISM) of gas and dust. The interstellar medium consists of 10^{-4} to 10^6 particles per cm^3 and is typically composed of roughly 70% hydrogen by mass, with most of the remaining gas consisting of helium. This medium has been chemically enriched by trace amounts of heavier elements that were ejected from stars as they passed beyond the end of their main sequence lifetime. Higher density regions of the interstellar medium form clouds, or *diffuse nebulae*, where star formation takes place. In contrast to spirals, an elliptical galaxy loses the cold component of its interstellar medium within roughly a billion years, which hinders the galaxy from forming diffuse nebulae except through mergers with other galaxies.

In the dense nebulae where stars are produced, much of the hydrogen is in the molecular (H_2) form, so these nebulae are called molecular clouds. Observations indicate that the coldest clouds tend to form low-mass stars, observed first in the infrared inside the clouds, then in visible light at their surface when the clouds dissipate, while giant molecular clouds, which are generally warmer, produce stars of all masses. These giant molecular clouds have typical densities of 100 particles per cm^3, diameters of 100 light-years (9.5×10^{14} km), masses of up to 6 million solar masses ($M_\odot$), and an average interior temperature of 10 K. About half the total mass of the galactic ISM is found in molecular clouds and in the Milky Way there are an estimated 6,000 molecular clouds, each with more than 100,000 $M_\odot$. The nearest nebula to the Sun where massive stars are being formed is the Orion nebula, 1,300 ly (1.2×10^{16} km) away. However, lower mass star formation is occurring about 400–450 light years distant in the ρ Ophiuchi cloud complex.

A more compact site of star formation is the opaque clouds of dense gas and dust known as Bok globules; so named after the astronomer Bart Bok. These can form in association with collapsing molecular clouds or possibly independently. The Bok globules are typically up to a light year across and contain a few solar masses. They can be observed as dark clouds silhouetted against bright emission nebulae or background stars. Over half the known Bok globules have been found to contain newly forming stars.

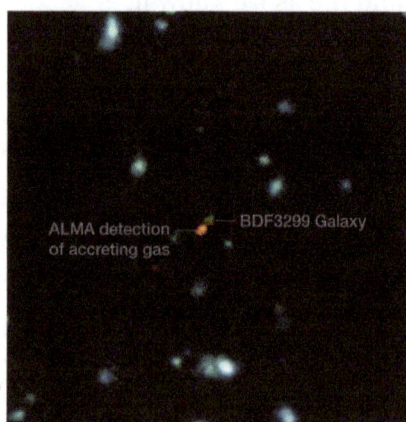

Assembly of galaxy in early Universe.

Cloud Collapse

An interstellar cloud of gas will remain in hydrostatic equilibrium as long as the kinetic energy of the gas pressure is in balance with the potential energy of the internal gravitational force. Mathematically this is expressed using the virial theorem, which states that, to maintain equilibrium, the gravitational potential energy must equal twice the internal thermal energy. If a cloud is massive enough that the gas pressure is insufficient to support it, the cloud will undergo gravitational collapse. The mass above which a cloud will undergo such collapse is called the Jeans mass. The Jeans mass depends on the temperature and density of the cloud, but is typically thousands to tens of thousands of solar masses. This coincides with the typical mass of an open cluster of stars, which is the end product of a collapsing cloud.

Stellar cluster and star-forming region M 17.

In *triggered star formation*, one of several events might occur to compress a molecular cloud and initiate its gravitational collapse. Molecular clouds may collide with each other, or a nearby supernova explosion can be a trigger, sending shocked matter into the cloud at very high speeds. Alternatively, galactic collisions can trigger massive starbursts of star formation as the gas clouds in each galaxy are compressed and agitated by tidal forces. The latter mechanism may be responsible for the formation of globular clusters.

A supermassive black hole at the core of a galaxy may serve to regulate the rate of star formation in a galactic nucleus. A black hole that is accreting infalling matter can become active, emitting a strong wind through a collimated relativistic jet. This can limit further star formation. Massive black holes ejecting radio-frequency-emitting particles at near-light speed can also block the formation of new stars in aging galaxies. However, the radio emissions around the jets may also trigger star formation. Likewise, a weaker jet may trigger star formation when it collides with a cloud.

As it collapses, a molecular cloud breaks into smaller and smaller pieces in a hierarchical manner, until the fragments reach stellar mass. In each of these fragments, the collapsing gas radiates away the energy gained by the release of gravitational potential energy. As the density increases, the fragments become opaque and are thus less efficient at radiating away their energy. This raises the temperature of the cloud and inhibits further fragmentation. The fragments now condense into rotating spheres of gas that serve as stellar embryos.

Complicating this picture of a collapsing cloud are the effects of turbulence, macroscopic flows, rotation, magnetic fields and the cloud geometry. Both rotation and magnetic fields can hinder the collapse of a cloud. Turbulence is instrumental in causing fragmentation of the cloud, and on the smallest scales it promotes collapse.

Protostar

LH 95 stellar nursery in Large Magellanic Cloud.

A protostellar cloud will continue to collapse as long as the gravitational binding energy can be eliminated. This excess energy is primarily lost through radiation. However, the collapsing cloud will eventually become opaque to its own radiation, and the energy must be removed through some other means. The dust within the cloud becomes heated to temperatures of 60–100 K, and these particles radiate at wavelengths in the far infrared where the cloud is transparent. Thus the dust mediates the further collapse of the cloud.

During the collapse, the density of the cloud increases toward the center and thus the middle region becomes optically opaque first. This occurs when the density is about 10^{-13} g / cm³. A core region, called the First Hydrostatic Core, forms where the collapse is essentially halted. It continues to increase in temperature as determined by the virial theorem. The gas falling toward this opaque region collides with it and creates shock waves that further heat the core.

Composite image showing young stars in and around molecular cloud Cepheus B.

When the core temperature reaches about 2000 K, the thermal energy dissociates the H_2 molecules. This is followed by the ionization of the hydrogen and helium atoms. These processes absorb the energy of the contraction, allowing it to continue on timescales comparable to the period of collapse at free fall velocities. After the density of infalling material has reached about 10^{-8} g / cm³, that material is sufficiently transparent to allow energy radiated by the protostar to escape. The combination of convection within the protostar and radiation from its exterior allow the star to contract further. This continues until the gas is hot enough for the internal pressure to support the protostar against further gravitational collapse—a state called hydrostatic equilibrium. When this accretion phase is nearly complete, the resulting object is known as a protostar.

N11, part of a complex network of gas clouds and star clusters within our neighbouring galaxy, the Large Magellanic Cloud.

Accretion of material onto the protostar continues partially from the newly formed circumstellar disc. When the density and temperature are high enough, deuterium fusion begins, and the outward pressure of the resultant radiation slows (but does not stop) the collapse. Material comprising the cloud continues to "rain" onto the protostar. In this stage bipolar jets are produced called Herbig-Haro objects. This is probably the means by which excess angular momentum of the infalling material is expelled, allowing the star to continue to form.

When the surrounding gas and dust envelope disperses and accretion process stops, the star is considered a pre–main sequence star (PMS star). The energy source of these objects is gravitational contraction, as opposed to hydrogen burning in main sequence stars. The PMS star follows a Hayashi track on the Hertzsprung–Russell (H–R) diagram. The contraction will proceed until the Hayashi limit is reached, and thereafter contraction will continue on a Kelvin–Helmholtz timescale with the temperature remaining stable. Stars with less than $0.5\ M_{\odot}$ thereafter join the main sequence. For more massive PMS stars, at the end of the Hayashi track they will slowly collapse in near hydrostatic equilibrium, following the Henyey track.

Finally, hydrogen begins to fuse in the core of the star, and the rest of the enveloping material is cleared away. This ends the protostellar phase and begins the star's main sequence phase on the H–R diagram.

The stages of the process are well defined in stars with masses around $1\ M_{\odot}$ or less. In high mass stars, the length of the star formation process is comparable to the other timescales of their evolution, much shorter, and the process is not so well defined. The later evolution of stars are studied in stellar evolution.

Protostar

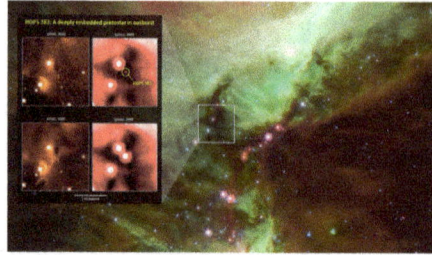

Protostar outburst - HOPS 383 (2015).

Observations

The Orion Nebula is an archetypical example of star formation, from the massive, young stars that are shaping the nebula to the pillars of dense gas that may be the homes of budding stars.

Key elements of star formation are only available by observing in wavelengths other than the optical. The protostellar stage of stellar existence is almost invariably hidden away deep inside dense clouds of gas and dust left over from the GMC. Often, these star-forming cocoons known as Bok globules, can be seen in silhouette against bright emission from surrounding gas. Early stages of a star's life can be seen in infrared light, which penetrates the dust more easily than visible light. Observations from the Wide-field Infrared Survey Explorer (WISE) have thus been especially important for unveiling numerous Galactic protostars and their parent star clusters. Examples of such embedded star clusters are FSR 1184, FSR 1190, Camargo 14, Camargo 74, Majaess 64, and Majaess 98.

Star-forming Region S106.

The structure of the molecular cloud and the effects of the protostar can be observed in near-IR extinction maps (where the number of stars are counted per unit area and compared to a nearby zero extinction area of sky), continuum dust emission and rotational transitions of CO and other molecules; these last two are observed in the millimeter and submillimeter range. The radiation from the protostar and early star has to be observed in infrared astronomy wavelengths, as the extinction caused by the rest of the cloud in which the star is forming is usually too big to allow us to observe it in the visual part of the spectrum. This presents considerable difficulties as the Earth's atmosphere is almost entirely opaque from 20μm to 850μm, with narrow windows at 200μm and 450μm. Even outside this range, atmospheric subtraction techniques must be used.

Young stars (purple) revealed by X-ray inside the NGC 2024 star-forming region.

X-ray observations have proven useful for studying young stars, since X-ray emission from these objects is 100–100,000 times stronger than X-ray emission from main-sequence stars. The earliest detections of X-rays from T Tauri stars were made by the Einstein X-ray Observatory. For low-mass stars X-rays are generated by the heating of the stellar corona through magnetic reconnection, while for high-mass O and early B-type stars X-rays are generated through supersonic shocks in the stellar winds. Photons in the soft X-ray energy range covered by the Chandra X-ray Observatory and XMM Newton may penetrate the interstellar medium with only moderate absorption due to gas, making the X-ray a useful wavelength for seeing the stellar populations within molecular clouds. X-ray emission as evidence of stellar youth makes this band particularly useful for performing censuses of stars in star-forming regions, given that not all young stars have infrared excesses. X-ray observations have provided near-complete censuses of all stellar-mass objects in the Orion Nebula Cluster and Taurus Molecular Cloud.

The formation of individual stars can only be directly observed in the Milky Way Galaxy, but in distant galaxies star formation has been detected through its unique spectral signature.

The first observed newborn star-forming clump, aged less than 10 million years old, was found in a galaxy about a corresponding light travel distance of 10.4 billion light years away, at an age when the universe was about 3.3 billion years old. The clump is about 3,000 light-years wide, and has a mass more than 1 billion times the mass of the sun, creating 32 stars each year with the mass of the sun, and produced about 40 percent of the stars in the clump's host galaxy.

Initial research indicates star-forming clumps start as giant, dense areas in turbulent gas-rich matter in young galaxies, live about 500 million years, and may migrate to the center of a galaxy, creating the central bulge of a galaxy.

On February 21, 2014, NASA announced a greatly upgraded database for tracking polycyclic aromatic hydrocarbons (PAHs) in the universe. According to scientists, more than 20% of the carbon in the universe may be associated with PAHs, possible starting materials for the formation of life. PAHs seem to have been formed shortly after the Big Bang, are widespread throughout the universe, and are associated with new stars and exoplanets.

Notable pathfinder objects

- MWC 349 was first discovered in 1978, and is estimated to be only 1,000 years old.

- VLA 1623 – The first exemplar Class 0 protostar, a type of embedded protostar that has yet to accrete the majority of its mass. Found in 1993, is possibly younger than 10,000 years .

- L1014 – An incredibly faint embedded object representative of a new class of sources that are only now being detected with the newest telescopes. Their status is still undetermined, they could be the youngest low-mass Class 0 protostars yet seen or even very low-mass evolved objects (like a brown dwarf or even an interstellar planet).

- IRS 8* – The youngest known main sequence star in the Galactic Center region, discovered in August 2006. It is estimated to be 3.5 million years old .

Low mass and high mass star formation

Star-forming region Westerhout 40 and the Serpens-Aquila Rift- cloud filaments containing new stars fill the region.

Stars of different masses are thought to form by slightly different mechanisms. The theory of low-mass star formation, which is well-supported by a plethora of observations, suggests that low-mass stars form by the gravitational collapse of rotating density enhancements within molecular clouds. As described above, the collapse of a rotating cloud of gas and dust leads to the formation of an accretion disk through which matter is channeled onto a central protostar. For stars with masses higher than about 8 $M_\odot$, however, the mechanism of star formation is not well understood.

Massive stars emit copious quantities of radiation which pushes against infalling material. In the past, it was thought that this radiation pressure might be substantial enough to halt accretion onto the massive protostar and prevent the formation of stars with masses more than a few tens of solar masses. Recent theoretical work has shown that the production of a jet and outflow clears a cavity through which much of the radiation from a massive protostar can escape without hindering accretion through the disk and onto the protostar. Present thinking is that massive stars may therefore be able to form by a mechanism similar to that by which low mass stars form.

There is mounting evidence that at least some massive protostars are indeed surrounded by accretion disks. Several other theories of massive star formation remain to be tested observationally. Of these, perhaps the most prominent is the theory of competitive accretion, which suggests that massive protostars are "seeded" by low-mass protostars which compete with other protostars to draw in matter from the entire parent molecular cloud, instead of simply from a small local region.

Another theory of massive star formation suggests that massive stars may form by the coalescence of two or more stars of lower mass.

Stellar Evolution

Mass (solar masses)	Time (years)	Spectral type
60	3 million	O3
30	11 million	O7
10	32 million	B4
3	370 million	A5
1.5	3 billion	F5
1	10 billion	G2 (Sun)
0.1	1000s billions	M7

Representative lifetimes of stars as a function of their masses

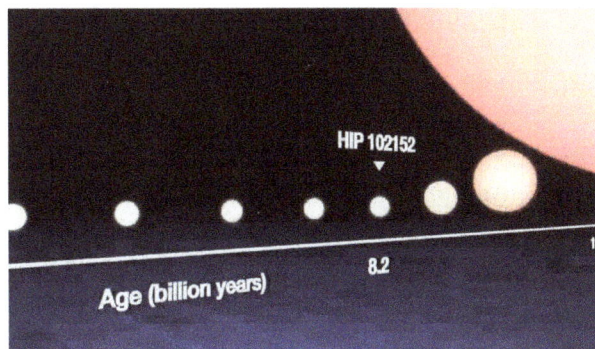

The life cycle of a Sun-like star.

Artist's depiction of the life cycle of a Sun-like star, starting as a main-sequence star at lower left then expanding through the subgiant and giant phases, until its outer envelope is expelled to form a planetary nebula at upper right.

Stellar evolution is the process by which a star changes over the course of time. Depending on the mass of the star, its lifetime can range from a few million years for the most massive to trillions of years for the least massive, which is considerably longer than the age of the universe. The table shows the lifetimes of stars as a function of their masses. All stars are born from collapsing clouds of gas and dust, often called nebulae or molecular clouds. Over the course of millions of years, these protostars settle down into a state of equilibrium, becoming what is known as a main-sequence star.

Nuclear fusion powers a star for most of its life. Initially the energy is generated by the fusion of hydrogen atoms at the core of the main-sequence star. Later, as the preponderance of atoms at the core becomes helium, stars like the Sun begin to fuse hydrogen along a spherical shell surrounding the core. This process causes the star to gradually grow in size, passing through the subgiant stage until it reaches the red giant phase. Stars with at least half the mass of the Sun can also begin to generate energy through the fusion of helium at their core, whereas more-massive stars can fuse heavier elements along a series of concentric shells. Once a star like the Sun has exhausted its nuclear fuel, its core collapses into a dense white dwarf and the outer layers are expelled as a planetary nebula. Stars with around ten or more times the mass of the Sun can explode in a supernova as their inert iron cores collapse into an extremely dense neutron star or black hole. Although the universe is not old enough for any of the smallest red dwarfs to have reached the end of their lives, stellar models suggest they will slowly become brighter and hotter before running out of hydrogen fuel and becoming low-mass white dwarfs.

Stellar evolution is not studied by observing the life of a single star, as most stellar changes occur too slowly to be detected, even over many centuries. Instead, astrophysicists come to understand how stars evolve by observing numerous stars at various points in their lifetime, and by simulating stellar structure using computer models.

In June 2015, astronomers reported evidence for Population III stars in the Cosmos Redshift 7 galaxy at $z = 6.60$. Such stars are likely to have existed in the very early universe (i.e., at high redshift), and may have started the production of chemical elements heavier than hydrogen that are needed for the later formation of planets and life as we know it.

Birth of a star

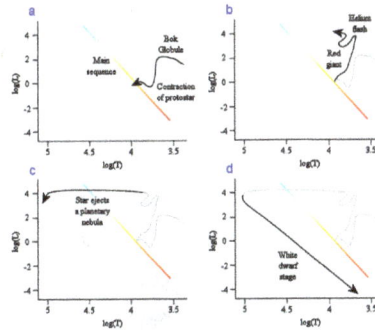

Schematic of stellar evolution.

Protostar

Stellar evolution starts with the gravitational collapse of a giant molecular cloud. Typical giant molecular clouds are roughly 100 light-years (9.5×10^{14} km) across and contain up to 6,000,000 solar masses (1.2×10^{37} kg). As it collapses, a giant molecular cloud breaks into smaller and smaller pieces. In each of these fragments, the collapsing gas releases gravitational potential energy as heat. As its temperature and pressure increase, a fragment condenses into a rotating sphere of superhot gas known as a protostar.

A protostar continues to grow by accretion of gas and dust from the molecular cloud, becoming a pre-main-sequence star as it reaches its final mass. Further development is determined by its mass. (Mass is compared to the mass of the Sun: 1.0 $M_\odot$ (2.0×10^{30} kg) means 1 solar mass.)

Protostars are encompassed in dust, and are thus more readily visible at infrared wavelengths. Observations from the Wide-field Infrared Survey Explorer (WISE) have been especially important for unveiling numerous Galactic protostars and their parent star clusters.

Brown dwarfs and sub-stellar objects

Protostars with masses less than roughly 0.08 $M_\odot$ (1.6×10^{29} kg) never reach temperatures high enough for nuclear fusion of hydrogen to begin. These are known as brown dwarfs. The International Astronomical Union defines brown dwarfs as stars massive enough to fuse deuterium at some point in their lives (13 Jupiter masses (M_J), 2.5×10^{28} kg, or 0.0125 $M_\odot$). Objects smaller than 13 M_J are classified as sub-brown dwarfs (but if they orbit around another stellar object they are classified as planets). Both types, deuterium-burning and not, shine dimly and die away slowly, cooling gradually over hundreds of millions of years.

For a more-massive protostar, the core temperature will eventually reach 10 million kelvin, initiating the proton–proton chain reaction and allowing hydrogen to fuse, first to deuterium and then to helium. In stars of slightly over 1 $M_\odot$ (2.0×10^{30} kg), the carbon–nitrogen–oxygen fusion reaction (CNO cycle) contributes a large portion of the energy generation. The onset of nuclear fusion leads relatively quickly to a hydrostatic equilibrium in which energy released by the core exerts a "radiation pressure" balancing the weight of the star's matter, preventing further gravitational collapse. The star thus evolves rapidly to a stable state, beginning the main-sequence phase of its evolution.

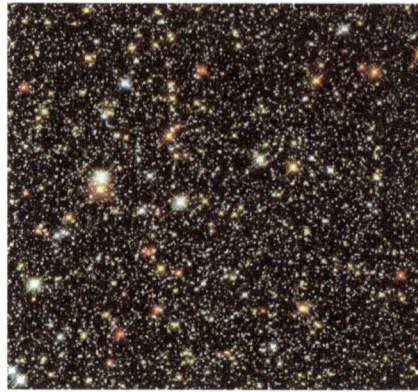

A dense starfield in Sagittarius

A new star will sit at a specific point on the main sequence of the Hertzsprung–Russell diagram, with the main-sequence spectral type depending upon the mass of the star. Small, relatively cold, low-mass red dwarfs fuse hydrogen slowly and will remain on the main sequence for hundreds of billions of years or longer, whereas massive, hot O-type stars will leave the main sequence after just a few million years. A mid-sized yellow dwarf star, like the Sun, will remain on the main sequence for about 10 billion years. The Sun is thought to be in the middle of its main sequence lifespan.

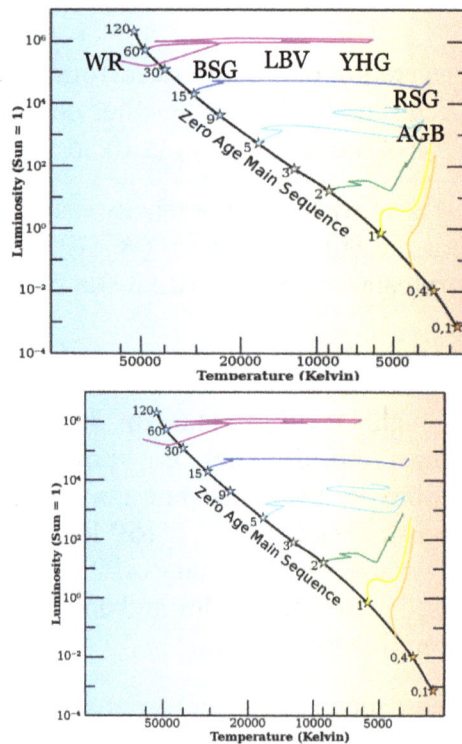

The evolutionary tracks of stars with different initial masses on the Hertzsprung–Russell diagram. The tracks start once the star has evolved to the main sequence and stop when fusion stops (for massive stars) and at the end of the red giant branch (for stars $1\,M_\odot$ and less).

A yellow track is shown for the Sun, which will become a red giant after its main-sequence phase ends before expanding further along the asymptotic giant branch, which will be the last phase in which the Sun undergoes fusion.

Mature stars

Eventually the core exhausts its supply of hydrogen and the star begins to evolve off of the main sequence. Without the outward pressure generated by the fusion of hydrogen to counteract the force of gravity the core contracts until either electron degeneracy pressure becomes sufficient to oppose gravity or the core becomes hot enough (around 100 MK) for helium fusion to begin. Which of these happens first depends upon the star's mass.

Low-mass stars

What happens after a low-mass star ceases to produce energy through fusion has not been directly observed; the universe is around 13.8 billion years old, which is less time (by several orders of magnitude, in some cases) than it takes for fusion to cease in such stars.

Recent astrophysical models suggest that red dwarfs of 0.1 $M_\odot$ may stay on the main sequence for some six to twelve trillion years, gradually increasing in both temperature and luminosity, and take several hundred billion more to collapse, slowly, into a white dwarf. Such stars will not become red giants as they are fully convective and will not develop a degenerate helium core with a shell burning hydrogen. Instead, hydrogen fusion will proceed until almost the whole star is helium.

Slightly more massive stars do expand into red giants, but their helium cores are not massive enough to reach the temperatures required for helium fusion so they never reach the tip of the red giant branch. When hydrogen shell burning finishes, these stars move directly off the red giant branch like a post AGB star, but at lower luminosity, to become a white dwarf. A star of about 0.5 $M_\odot$ will be able to reach temperatures high enough to fuse helium, and these "mid-sized" stars go on to further stages of evolution beyond the red giant branch.

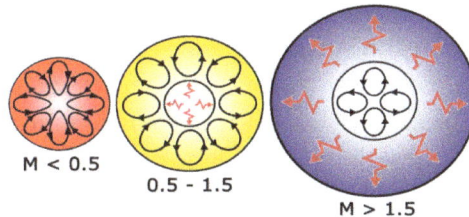

Internal structures of main-sequence stars, convection zones with arrowed cycles and radiative zones with red flashes. To the left a **low-mass** red dwarf, in the center a **mid-sized** yellow dwarf and at the right a **massive** blue-white main-sequence star.

Mid-sized Stars

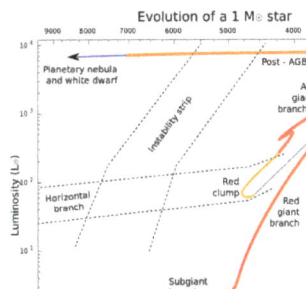

The evolutionary track of a solar mass, solar metallicity, star from main sequence to post-AGB

Stars of roughly 0.5–10 $M_\odot$ become red giants, which are large non-main-sequence stars of stellar classification K or M. Red giants lie along the right edge of the Hertzsprung–Russell diagram due to their red color and large luminosity. Examples include Aldebaran in the constellation Taurus and Arcturus in the constellation of Boötes.

Mid-sized stars are red giants during two different phases of their post-main-sequence evolution: red-giant-branch stars, whose inert cores are made of helium, and asymptotic-giant-branch stars, whose inert cores are made of carbon. Asymptotic-giant-branch stars have helium-burning shells inside the hydrogen-burning shells, whereas red-giant-branch stars have hydrogen-burning shells only. Between these two phases, stars spend a period on the horizontal branch with a helium-fusing core. Many of these helium-fusing stars cluster towards the cool end of the horizontal branch as K-type giants and are referred to red clump giants.

Subgiant Phase

When a star exhausts the hydrogen in its core, it leaves the main sequence and begins to fuse hydrogen in a shell outside the core. The core increases in mass as the shell produces more helium. Depending on the mass of the helium core, this continues for several million to one or two billion years, with the star expanding and cooling at a similar or slightly lower luminosity to its main sequence state. Eventually either the core becomes degenerate, in stars around the mass of the sun, or the outer layers cool sufficiently to become opaque, in more massive stars. Either of these changes cause the hydrogen shell to increase in temperature and the luminosity of the star to increase, at which point the star expands onto the red giant branch.

Red-giant-branch Phase

The expanding outer layers of the star are convective, with the material being mixed by turbulence from near the fusing regions up to the surface of the star. For all but the lowest-mass stars, the fused material has remained deep in the stellar interior prior to this point, so the convecting envelope makes fusion products visible at the star's surface for the first time. At this stage of evolution, the results are subtle, with the largest effects, alterations to the isotopes of hydrogen and helium, being unobservable. The effects of the CNO cycle appear at the surface during the first dredge-up, with lower $^{12}C/^{13}C$ ratios and altered proportions of carbon and nitrogen. These are detectable with spectroscopy and have been measured for many evolved stars.

The helium core continues to grow on the red giant branch. It is no longer in thermal equilibrium, either degenerate or above the Schoenberg-Chandrasekhar limit, so it increases in temperature which causes the rate of fusion in the hydrogen shell to increase. The star increases in luminosity towards the tip of the red-giant branch. Red giant branch stars with a degenerate helium core all reach the tip with very similar core masses and very similar luminosities, although the more massive of the red giants become hot enough to ignite helium fusion before that point.

Horizontal Branch

If the core is largely supported by electron degeneracy pressure, helium fusion will ignite on a timescale of days in a helium flash. In more massive stars, the ignition of helium fusion occurs relatively slowly with no flash. The nuclear power released during the helium flash is very large, on the order of 10^8 times the luminosity of the Sun for a few days and 10^{11} times the luminosity of the Sun (roughly the luminosity of the Milky Way Galaxy) for a few seconds. However, the energy is absorbed by the stellar envelope and thus cannot be seen from outside the star. The energy released by helium fusion causes the core to expand, so that hydrogen fusion in the overlying layers slows and total energy generation decreases. The star contracts, although not all the way to the main sequence, and it migrates to the horizontal branch on the Hertzsprung–Russell diagram, gradually shrinking in radius and increasing its surface temperature.

Core helium flash stars evolve to the red end of the horizontal branch but do not migrate to higher temperatures before they gain a degenerate carbon-oxygen core and start helium shell burning. These stars are often observed as a red clump of stars in the colour-magnitude diagram of a cluster, hotter and less luminous than the red giants. Higher-mass stars with larger helium cores move along the horizontal branch to higher temperatures, some becoming unstable pulsating stars in the yellow instability strip (RR Lyrae variables), whereas some become even hotter and can form a blue tail or blue hook to the horizontal branch. The exact morphology of the horizontal branch depends on parameters such as metallicity, age, and helium content, but the exact details are still being modelled.

Asymptotic-giant-branch Phase

After a star has consumed the helium at the core, hydrogen and helium fusion continues in shells around a hot core of carbon and oxygen. The star follows the asymptotic giant branch on the Hertzsprung–Russell diagram, paralleling the original red giant evolution, but with even faster energy generation (which lasts for a shorter time). Although helium is being burnt in a shell, the majority of the energy is produced by hydrogen burning in a shell further from the core of the star. Helium from these hydrogen burning shells drops towards the center of the star and periodically the energy output from the helium shell increases dramatically. This is known as a thermal pulse and they occur towards the end of the asymptotic-giant-branch phase, sometimes even into the post-asymptotic-giant-branch phase. Depending on mass and composition, there may be several to hundreds of thermal pulses.

There is a phase on the ascent of the asymptotic-giant-branch where a deep convective zone forms and can bring carbon from the core to the surface. This is known as the second dredge up, and in some stars there may even be a third dredge up. In this way a carbon star is formed, very cool and strongly reddened stars showing strong carbon lines in their spectra. A process known as hot bottom burning may convert carbon into oxygen and nitrogen before it can be dredged to the surface, and the interaction between these processes determines the observed luminosities and spectra of carbon stars in particular clusters.

Another well known class of asymptotic-giant-branch stars are the Mira variables, which pulsate with well-defined periods of tens to hundreds of days and large amplitudes up to about 10 magnitudes (in the visual, total luminosity changes by a much smaller amount). In more-massive stars the stars become more luminous and the pulsation period is longer, leading to enhanced mass loss, and the stars become heavily obscured at visual wavelengths. These stars can be observed as OH/IR stars, pulsating in the infra-red and showing OH maser activity. These stars are clearly oxygen rich, in contrast to the carbon stars, but both must be produced by dredge ups.

Post-AGB

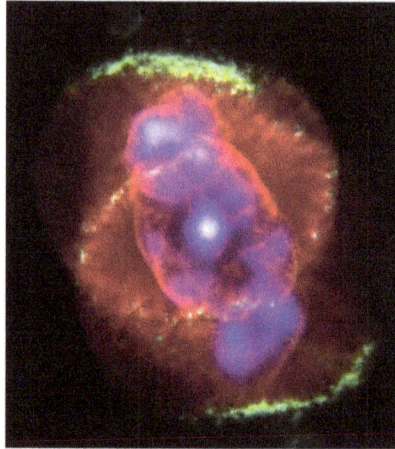

The Cat's Eye Nebula, a planetary nebula formed by the death of a star with about the same mass as the Sun

These mid-range stars ultimately reach the tip of the asymptotic-giant-branch and run out of fuel for shell burning. They are not sufficiently massive to start full-scale carbon fusion, so they contract again, going through a period of post-asymptotic-giant-branch superwind to produce a planetary nebula with an extremely hot central star. The central star then cools to a white dwarf. The expelled gas is relatively rich in heavy elements created within the star and may be particularly oxygen or carbon enriched, depending on the type of the star. The gas builds up in an expanding shell called a circumstellar envelope and cools as it moves away from the star, allowing dust particles and molecules to form. With the high infrared energy input from the central star, ideal conditions are formed in these circumstellar envelopes for maser excitation.

It is possible for thermal pulses to be produced once post-asymptotic-giant-branch evolution has begun, producing a variety of unusual and poorly understood stars known as born-again asymptotic-giant-branch stars. These may result in extreme horizontal-branch stars (subdwarf B stars), hydrogen deficient post-asymptotic-giant-branch stars, variable planetary nebula central stars, and R Coronae Borealis variables.

Massive Stars

The Crab Nebula, the shattered remnants of a star which exploded as a supernova, the light of which reached Earth in 1054 AD

In massive stars, the core is already large enough at the onset of the hydrogen burning shell that helium ignition will occur before electron degeneracy pressure has a chance to become prevalent. Thus, when these stars expand and cool, they do not brighten as much as lower-mass stars; however, they were much brighter than lower-mass stars to begin with, and are thus still brighter than the red giants formed from less-massive stars. These stars are unlikely to survive as red supergiants; instead they will destroy themselves as type II supernovas.

Extremely massive stars (more than approximately 40 $M_\odot$), which are very luminous and thus have very rapid stellar winds, lose mass so rapidly due to radiation pressure that they tend to strip off their own envelopes before they can expand to become red supergiants, and thus retain extremely high surface temperatures (and blue-white color) from their main-sequence time onwards. The largest stars of the current generation are about 100-150 $M_\odot$ because the outer layers would be expelled by the extreme radiation. Although lower-mass stars normally do not burn off their outer layers so rapidly, they can likewise avoid becoming red giants or red supergiants if they are in binary systems close enough so that the companion star strips off the envelope as it expands, or if they rotate rapidly enough so that convection extends all the way from the core to the surface, resulting in the absence of a separate core and envelope due to thorough mixing.

The core grows hotter and denser as it gains material from fusion of hydrogen at the base of the envelope. In all massive stars, electron degeneracy pressure is insufficient to halt collapse by itself, so as each major element is consumed in the center, progressively heavier elements ignite, temporarily halting collapse. If the core of the star is not too massive (less than approximately 1.4 $M_\odot$, taking into account mass loss that has occurred by this time), it may then form a white dwarf (possibly surrounded by a planetary nebula) as described above for less-massive stars, with the difference that the white dwarf is composed chiefly of oxygen, neon, and magnesium.

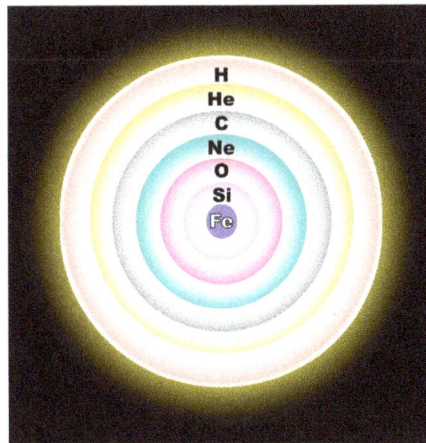

The onion-like layers of a massive, evolved star just before core collapse. (Not to scale.)

Above a certain mass (estimated at approximately 2.5 $M_\odot$ and whose star's progenitor was around 10 $M_\odot$), the core will reach the temperature (approximately 1.1 gigakelvins) at which neon partially breaks down to form oxygen and helium, the latter of which immediately fuses with some of the remaining neon to form magnesium; then oxygen fuses to form sulfur, silicon, and smaller amounts of other elements. Finally, the temperature gets high enough that any nucleus can be partially broken down, most commonly releasing an alpha particle (helium nucleus) which immediately fuses with another nucleus, so that several nuclei are effectively rearranged into a smaller number of heavier nuclei, with net release of energy because the addition of fragments to nuclei exceeds the energy required to break them off the parent nuclei.

A star with a core mass too great to form a white dwarf but insufficient to achieve sustained conversion of neon to oxygen and magnesium, will undergo core collapse (due to electron capture) before achieving fusion of the heavier elements. Both heating and cooling caused by electron capture onto minor constituent elements (such as aluminum and sodium) prior to collapse may have a significant impact on total energy generation within the star shortly before collapse. This may produce a noticeable effect on the abundance of elements and isotopes ejected in the subsequent supernova.

Supernova

Once the nucleosynthesis process arrives at iron-56, the continuation of this process consumes energy (the addition of fragments to nuclei releases less energy than required to break them off the parent nuclei). If the mass of the core exceeds the Chandrasekhar limit, electron degeneracy pressure will be unable to support its weight against the force of gravity, and the core will undergo sudden, catastrophic collapse to form a neutron star or (in the case of cores that exceed the Tolman-Oppenheimer-Volkoff limit), a black hole. Through a process that is not completely understood, some of the gravitational potential energy released by this core collapse is converted into a Type Ib, Type Ic, or Type II supernova. It is known that the core collapse produces a massive surge of neutrinos, as observed with supernova SN 1987A. The extremely energetic neutrinos fragment some nuclei; some of their energy is consumed in releasing nucleons, including neutrons, and some of their energy is transformed into heat and kinetic energy, thus augmenting the shock wave started by rebound of some of the infalling material from the collapse of the core. Electron capture in very dense parts of the infalling matter may produce additional neutrons. Because some of the rebounding matter is bombarded by the neutrons, some of its nuclei capture them, creating a spectrum of heavier-than-iron material including the radioactive elements up to (and likely beyond) uranium. Although non-exploding red giants can produce significant quantities of elements heavier than iron using neutrons released in side reactions of earlier nuclear reactions, the abundance of elements heavier than iron (and in particular, of certain isotopes of elements that have multiple stable or long-lived isotopes) produced in such reactions is quite different from that produced in a supernova. Neither abundance alone matches that found in the Solar System, so both supernovae and ejection of elements from red giants are required to explain the observed abundance of heavy elements and isotopes thereof.

The energy transferred from collapse of the core to rebounding material not only generates heavy elements, but provides for their acceleration well beyond escape velocity, thus causing a Type Ib, Type Ic, or Type II supernova. Note that current understanding of this energy transfer is still not satisfactory; although current computer models of Type Ib, Type Ic, and Type II supernovae account for part of the energy transfer, they are not able to account for enough energy transfer to produce the observed ejection of material.

Some evidence gained from analysis of the mass and orbital parameters of binary neutron stars (which require two such supernovae) hints that the collapse of an oxygen-neon-magnesium core may produce a supernova that differs observably (in ways other than size) from a supernova produced by the collapse of an iron core.

The most-massive stars that exist today may be completely destroyed by a supernova with an energy greatly exceeding its gravitational binding energy. This rare event, caused by pair-instability, leaves behind no black hole remnant. In the past history of the universe, some stars were even larger than the largest that exists today, and they would immediately collapse into a black hole at the end of their lives, due to photodisintegration.

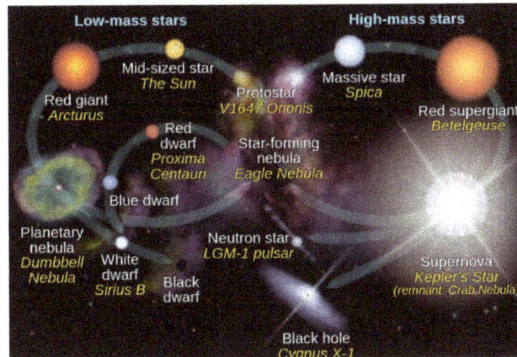

Stellar evolution of low-mass (left cycle) and high-mass (right cycle) stars, with examples in italics

Stellar Remnants

After a star has burned out its fuel supply, its remnants can take one of three forms, depending on the mass during its lifetime.

White and Black Dwarfs

For a star of 1 $M_\odot$, the resulting white dwarf is of about 0.6 $M_\odot$, compressed into approximately the volume of the Earth. White dwarfs are stable because the inward pull of gravity is balanced by the degeneracy pressure of the star's electrons, a consequence of the Pauli exclusion principle. Electron degeneracy pressure provides a rather soft limit against further compression; therefore, for a given chemical composition, white dwarfs of higher mass have a smaller volume. With no fuel left to burn, the star radiates its remaining heat into space for billions of years.

A white dwarf is very hot when it first forms, more than 100,000 K at the surface and even hotter in its interior. It is so hot that a lot of its energy is lost in the form of neutrinos for the first 10 million years of its existence, but will have lost most of its energy after a billion years.

The chemical composition of the white dwarf depends upon its mass. A star of a few solar masses will ignite carbon fusion to form magnesium, neon, and smaller amounts of other elements, resulting in a white dwarf composed chiefly of oxygen, neon, and magnesium, provided that it can lose enough mass to get below the Chandrasekhar limit, and provided that the ignition of carbon is not so violent as to blow the star apart in a supernova. A star of mass on the order of magnitude of the Sun will be unable to ignite carbon fusion, and will produce a white dwarf composed chiefly of carbon and oxygen, and of mass too low to collapse unless matter is added to it later. A star of less than about half the mass of the Sun will be unable to ignite helium fusion (as noted earlier), and will produce a white dwarf composed chiefly of helium.

In the end, all that remains is a cold dark mass sometimes called a black dwarf. However, the universe is not old enough for any black dwarfs to exist yet.

If the white dwarf's mass increases above the Chandrasekhar limit, which is 1.4 $M_\odot$ for a white dwarf composed chiefly of carbon, oxygen, neon, and/or magnesium, then electron degeneracy pressure fails due to electron capture and the star collapses. Depending upon the chemical composition and pre-collapse temperature in the center, this will lead either to collapse into a neutron star or runaway ignition of carbon and oxygen. Heavier elements favor continued core collapse, because they require a higher temperature to ignite, because electron capture onto these elements and their fusion products is easier; higher core temperatures favor runaway nuclear reaction, which halts core collapse and leads to a Type Ia supernova. These supernovae may be many times brighter than the Type II supernova marking the death of a massive star, even though the latter has the greater total energy release. This instability to collapse means that no white dwarf more massive than approximately 1.4 $M_\odot$ can exist (with a possible minor exception for very rapidly spinning white dwarfs, whose centrifugal force due to rotation partially counteracts the weight of their matter). Mass transfer in a binary system may cause an initially stable white dwarf to surpass the Chandrasekhar limit.

If a white dwarf forms a close binary system with another star, hydrogen from the larger companion may accrete around and onto a white dwarf until it gets hot enough to fuse in a runaway reaction at its surface, although the white dwarf remains below the Chandrasekhar limit. Such an explosion is termed a nova.

Neutron Stars

Bubble-like shock wave still expanding from a supernova explosion 15,000 years ago.

Ordinarily, atoms are mostly electron clouds by volume, with very compact nuclei at the center (proportionally, if atoms were the size of a football stadium, their nuclei would be the size of dust mites). When a stellar core collapses, the pressure causes electrons and protons to fuse by electron capture. Without electrons, which keep nuclei apart, the neutrons collapse into a dense ball (in some ways like a giant atomic nucleus), with a thin overlying layer of degenerate matter (chiefly iron unless matter of different composition is added later). The neutrons resist further compression by the Pauli Exclusion Principle, in a way analogous to electron degeneracy pressure, but stronger.

These stars, known as neutron stars, are extremely small—on the order of radius 10 km, no bigger than the size of a large city—and are phenomenally dense. Their period of rotation shortens dramatically as the stars shrink (due to conservation of angular momentum); observed rotational periods of neutron stars range from about 1.5 milliseconds (over 600 revolutions per second) to several seconds. When these rapidly rotating stars' magnetic poles are aligned with the Earth, we detect a pulse of radiation each revolution. Such neutron stars are called pulsars, and were the first neutron stars to be discovered. Though electromagnetic radiation detected from pulsars is most often in the form of radio waves, pulsars have also been detected at visible, X-ray, and gamma ray wavelengths.

Black Holes

If the mass of the stellar remnant is high enough, the neutron degeneracy pressure will be insufficient to prevent collapse below the Schwarzschild radius. The stellar remnant thus becomes a black hole. The mass at which this occurs is not known with certainty, but is currently estimated at between 2 and 3 $M_\odot$.

Black holes are predicted by the theory of general relativity. According to classical general relativity, no matter or information can flow from the interior of a black hole to an outside observer, although quantum effects may allow deviations from this strict rule. The existence of black holes in the universe is well supported, both theoretically and by astronomical observation.

Because the core-collapse supernova mechanism itself is imperfectly understood, it is still not known whether it is possible for a star to collapse directly to a black hole without producing a visible supernova, or whether some supernovae initially form unstable neutron stars which then collapse into black holes; the exact relation between the initial mass of the star and the final remnant is also not completely certain. Resolution of these uncertainties requires the analysis of more supernovae and supernova remnants.

Models

A stellar evolutionary model is a mathematical model that can be used to compute the evolutionary phases of a star from its formation until it becomes a remnant. The mass and chemical composition of the star are used as the inputs, and the luminosity and surface temperature are the only constraints. The model formulae are based upon the physical understanding of the star, usually under the assumption of hydrostatic equilibrium. Extensive computer calculations are then run to determine the changing state of the star over time, yielding a table of data that can be used to determine the evolutionary track of the star across the Hertzsprung–Russell diagram, along with other evolving properties. Accurate models can be used to estimate the current age of a star by comparing its physical properties with those of stars along a matching evolutionary track.

Supernova

SN 1994D (bright spot on the lower left), a type Ia supernova in the NGC 4526 galaxy

A supernova is an astronomical event that occurs during the last stellar evolutionary stages of a massive star's life, whose dramatic and catastrophic destruction is marked by one final titanic explosion. For a short time, this causes the sudden appearance of a 'new' bright star, before slowly fading from sight over several weeks or months.

Only three Milky Way naked-eye supernova events have been observed during the last thousand years, though many have been telescopically seen in other galaxies. The most recent directly observed supernova in the Milky Way was Kepler's Star in 1604 (SN 1604), but remnants of two more recent supernovae have been found retrospectively. Statistical observations of supernovae in other galaxies suggest they should occur on average about three times every century in the Milky Way, and that any galactic supernova would almost certainly be observable in modern astronomical equipment.

Supernovae are more energetic than novae. In Latin, *Nova* means "new", referring astronomically to what appears to be a temporary new bright star. Adding the prefix "super-" distinguishes supernovae from ordinary novae, which are far less luminous. The word *supernova* was coined by Walter Baade and Fritz Zwicky in 1931.

During maximum brightness, the total equivalent radiant energies produced by supernovae may briefly outshine an entire output of a typical galaxy and emit energies equal to that created over the lifetime of any solar-like star. Such extreme catastrophes may also expel much, if not all, of its stellar material away from the star, at velocities up to 30,000 km/s or 10% of the speed of light. This drives an expanding and fast-moving shock wave into the surrounding interstellar medium, and in turn, sweeping up an expanding shell of gas and dust, which is observed as a supernova remnant. Supernovae create, fuse and eject the bulk of the chemical elements produced by nucleosynthesis. Supernovae play a significant role in enriching the interstellar medium with the heavier atomic mass chemical elements. Furthermore, the expanding shock waves from supernova explosions can trigger the formation of new stars. Supernova remnants are expected to accelerate a large fraction of galactic primary cosmic rays, but direct evidence for cos-mic ray production was found only in a few of them so far. They are also potentially strong galactic sources of gravitational waves.

Theoretical studies of many supernovae indicate that most are triggered by one of two basic mechanisms: the sudden re-ignition of nuclear fusion in a degenerate star or the sudden gravitational collapse of a massive star's core. In the first instance, a degenerate white dwarf may accumulate sufficient material from a binary companion, either through accretion or via a merger, to raise its core temperature enough to trigger runaway nuclear fusion, completely disrupting the star. In the second case, the core of a massive star may undergo sudden gravitational collapse, releasing gravitational potential energy as a supernova. While some observed supernovae are more complex than these two simplified theories, the astrophysical collapse mechanics have been established and accepted by most astronomers for some time.

Due to the wide range of astrophysical consequences of these events, astronomers now deem supernovae research, across the fields of stellar and galactic evolution, as an especially important area for investigation.

Etymology

In Latin, *Nova* means "new", referring astronomically to what appears to be a temporary new bright star. Adding the prefix "super-" distinguishes supernovae from ordinary novae, which are far less luminous. The word *supernova* was coined by Walter Baade and Fritz Zwicky in 1931.

Observation History

The Crab Nebula is a pulsar wind nebula associated with the 1054 supernova

The highlighted passages refer to the Chinese observation of SN 1054

The earliest recorded supernova, SN 185, was viewed by Chinese astronomers in 185 AD, with the brightest recorded supernova being SN 1006, which occurred in 1006 AD and was described in detail by Chinese and Islamic astronomers. The widely observed supernova SN 1054 produced the Crab Nebula. Supernovae SN 1572 and SN 1604, the latest to be observed with the naked eye in the Milky Way galaxy, had notable effects on the development of astronomy in Europe because they were used to argue against the Aristotelian idea that the universe beyond the Moon and planets was static and unchanging. Johannes Kepler began observing SN 1604 at its peak on October 17, 1604, and continued to make estimates of its brightness until it faded from naked eye view a year later. It was the second supernova to be observed in a generation (after SN 1572 seen by Tycho Brahe in Cassiopeia).

Before the development of the telescope, there have only been five supernovae seen in the last millennium. In the perspective of how long a star's lifetime is, its death is very brief. In fact, a star's death may only last a few months. Due to this, a typical human will only experience this rarity, on average, once in their lifetime. This is a microscopic fraction in comparison to the 100 billion stars that compose a galaxy. However, since the use of modern equipment, particularly in this millennium, professional and amateur astronomers have been finding several hundreds of supernovae each year.

The field of supernova discovery has extended to other galaxies, starting with SN 1885A in the Andromeda galaxy. American astronomers Rudolph Minkowski and Fritz Zwicky developed the modern supernova classification scheme beginning in 1941. In the 1960s, astronomers found that the maximum intensities of supernova explosions could be used as standard candles, hence indicators of astronomical distances. Some of the most distant supernovae recently observed appeared dimmer than expected. This supports the view that the expansion of the universe is accelerating. Techniques were developed for reconstructing supernova explosions that have no written records of being observed. The date of the Cassiopeia A supernova event was determined from light echoes off nebulae, while the age of supernova remnant RX J0852.0-4622 was estimated from temperature measurements and the gamma ray emissions from the decay of titanium-44.

The brightest observed supernova, ASASSN-15lh, was detected in June 2015. With a brightness of 570 billion Suns, ASASSN-15lh's peak luminosity was twice that of the previous record holder.

Discovery

Early work on what was originally believed to be simply a new category of novae was performed during the 1930s by Walter Baade and Fritz Zwicky at Mount Wilson Observatory. The name *super-novae* was first used during 1931 lectures held at Caltech by Baade and Zwicky, then used publicly in 1933 at a meeting of the American Physical Society. By 1938, the hyphen had been lost and the modern name was in use. Because supernovae are relatively rare events within a galaxy, occurring about three times a century in the Milky Way, obtaining a good sample of supernovae to study requires regular monitoring of many galaxies.

Supernovae in other galaxies cannot be predicted with any meaningful accuracy. Normally, when they are discovered, they are already in progress. Most scientific interest in supernovae—as standard candles for measuring distance, for example—require an observation of their peak luminosity. It is therefore important to discover them well before they reach their maximum. Amateur astronomers, who greatly outnumber professional astronomers, have played an important role in finding supernovae, typically by looking at some of the closer galaxies through an optical telescope and comparing them to earlier photographs.

A supernova remnant

Toward the end of the 20th century astronomers increasingly turned to computer-controlled telescopes and CCDs for hunting supernovae. While such systems are popular with amateurs, there are also professional installations such as the Katzman Automatic Imaging Telescope. Recently the Supernova Early Warning System (SNEWS) project has begun using a network of neutrino detectors to give early warning of a supernova in the Milky Way galaxy. Neutrinos are particles that are produced in great quantities by a supernova explosion, and they are not significantly absorbed by the interstellar gas and dust of the galactic disk.

"A star set to explode", the SBW1 nebula surrounds a massive blue supergiant in the Carina Nebula.

Supernova searches fall into two classes: those focused on relatively nearby events and those looking for explosions farther away. Because of the expansion of the universe, the distance to a remote object with a known emission spectrum can be estimated by measuring its Doppler shift (or redshift); on average, more distant objects recede with greater velocity than those nearby, and so have a higher redshift. Thus the search is split between high redshift and low redshift, with the boundary falling around a redshift range of $z = 0.1–0.3$—where z is a dimensionless measure of the spectrum's frequency shift.

High redshift searches for supernovae usually involve the observation of supernova light curves. These are useful for standard or calibrated candles to generate Hubble diagrams and make cosmological predictions. Supernova spectroscopy, used to study the physics and environments of supernovae, is more practical at low than at high redshift. Low redshift observations also anchor the low-distance end of the Hubble curve, which is a plot of distance versus redshift for visible galaxies.

Naming Convention

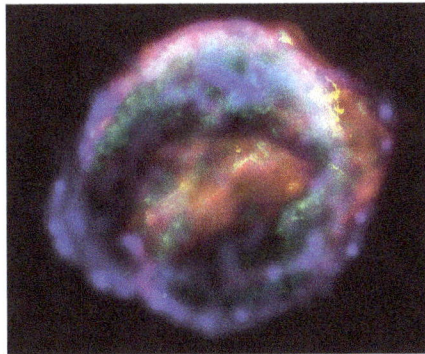

Multiwavelength X-ray, infrared, and optical compilation image of Kepler's supernova remnant, SN 1604.

Supernova discoveries are reported to the International Astronomical Union's Central Bureau for Astronomical Telegrams, which sends out a circular with the name it assigns to that supernova. The name is the marker *SN* followed by the year of discovery, suffixed with a one or two-letter designation. The first 26 supernovae of the year are designated with a capital letter from *A* to *Z*. Afterward pairs of lower-case letters are used: *aa*, *ab*, and so on. Hence, for example, *SN 2003C* designates the third supernova reported in the year 2003. The last supernova of 2005 was SN 2005nc, indicating that it was the 367th supernova found in 2005. Since 2000, professional and amateur astronomers have been finding several hundreds of supernovae each year (572 in 2007, 261 in 2008, 390 in 2009; 231 in 2013).

Historical supernovae are known simply by the year they occurred: SN 185, SN 1006, SN 1054, SN 1572 (called *Tycho's Nova*) and SN 1604 (*Kepler's Star*). Since 1885 the additional letter notation has been used, even if there was only one supernova discovered that year (e.g. SN 1885A, SN 1907A, etc.) — this last happened with SN 1947A. *SN*, for SuperNova, is a standard prefix. Until 1987, two-letter designations were rarely needed; since 1988, however, they have been needed every year.

Classification

As part of the attempt to understand supernovae, astronomers have classified them according to their light curves and the absorption lines of different chemical elements that appear in their spectra. The first element for division is the presence or absence of a line caused by hydrogen. If a supernova's spectrum contains lines of hydrogen (known as the Balmer series in the visual portion of the spectrum) it is classified *Type II*; otherwise it is *Type I*. In each of these two types there are subdivisions according to the presence of lines from other elements or the shape of the light curve (a graph of the supernova's apparent magnitude as a function of time).

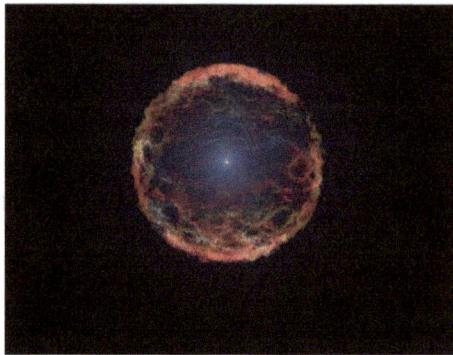

Artist's impression of supernova 1993J.

Supernova taxonomy					
Type I No hydrogen		Type Ia Presents a singly ionized silicon (Si II) line at 615.0 nm (nanometers), near peak light			Thermal runaway
	Type Ib/c Weak or no silicon absorption feature	Type Ib Shows a non-ionized helium (He I) line at 587.6 nm			Core collapse
		Type Ic Weak or no helium			
Type II Shows hydrogen	Type II-P/L/N Type II spectrum throughout	Type II-P/L No narrow lines	Type II-P Reaches a "plateau" in its light curve		
			Type II-L Displays a "linear" decrease in its light curve (linear in magnitude versus time).		
		Type IIn Some narrow lines			
	Type IIb Spectrum changes to become like Type Ib				

Type I

Type I supernovae are subdivided on the basis of their spectra, with type Ia showing a strong ionised silicon absorption line. Type I supernovae without this strong line are classified as Type Ib and Ic, with Type Ib showing strong neutral helium lines and Type Ic lacking them. The light curves are all similar, although Type Ia are generally brighter at peak luminosity, but the light curve is not important for classification of Type I supernovae.

A small number of Type Ia supernovae exhibit unusual features such as non-standard luminosity or broadened light curves, and these are typically classified by referring to the earliest example showing similar features. For example, the sub-luminous SN 2008ha is often referred to as SN 2002cx-like or class Ia-2002cx.

Type II

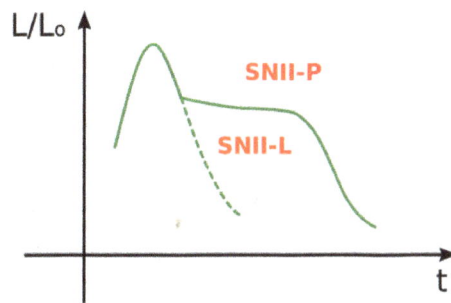

Light curves are used to classify type II-P and type II-L supernovae

The supernovae of Type II can also be sub-divided based on their spectra. While most Type II supernovae show very broad emission lines which indicate expansion velocities of many thousands of kilometres per second, some, such as SN 2005gl, have relatively narrow features in their spectra. These are called Type IIn, where the 'n' stands for 'narrow'.

A few supernovae, such as SN 1987K and SN 1993J, appear to change types: they show lines of hydrogen at early times, but, over a period of weeks to months, become dominated by lines of helium. The term "Type IIb" is used to describe the combination of features normally associated with Types II and Ib.

Type II supernovae with normal spectra dominated by broad hydrogen lines that remain for the life of the decline are classified on the basis of their light curves. The most common type shows a distinctive "plateau" in the light curve shortly after peak brightness where the visual luminosity stays relatively constant for several months before the decline resumes. These are called type II-P referring to the plateau. Less common are type II-L supernovae that lack a distinct plateau. The "L" signifies "linear" although the light curve is not actually a straight line.

Supernovae that do not fit into the normal classifications are designated peculiar, or 'pec'.

Types III, IV, and V

Fritz Zwicky defined additional supernovae types, although based on a very few examples that didn't cleanly fit the parameters for a Type I or Type II supernova. SN 1961i in NGC 4303 was the prototype and only member of the Type III supernova class, noted for its broad light curve maximum and broad hydrogen Balmer lines that were slow to develop in the spectrum. SN 1961f in NGC 3003 was the prototype and only member of the Type IV class, with a light curve similar to a Type II-P supernova, with hydrogen absorption lines but weak hydrogen emission lines. The Type V class was coined for SN 1961V in NGC 1058, an unusual faint supernova or supernova imposter with a slow rise to brightness, a maximum lasting many months, and an unusual emission spectrum. The similarity of SN 1961V to the Eta Carinae Great Outburst was noted. Supernovae in M101 (1909) and M83 (1923 and 1957) were also suggested as possible type IV or type V supernovae.

These types would now all be treated as peculiar Type II supernovae, of which many more examples have been discovered, although it is still debated whether SN 1961V was a true supernova following an LBV outburst or an imposter.

Current Models

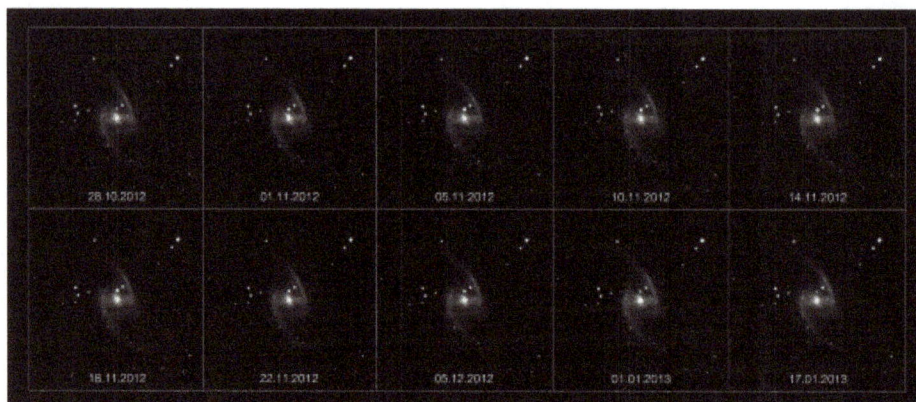

Sequence shows the rapid brightening and slower fading of a supernova explosion in the galaxy NGC 1365

The type codes, described above given to supernovae, are *taxonomic* in nature: the type number describes the light observed from the supernova, not necessarily its cause. For example, type Ia supernovae are produced by runaway fusion ignited on degenerate white dwarf progenitors while the spectrally similar type Ib/c are produced from massive Wolf-Rayet progenitors by core collapse. The following summarizes what is currently believe to be the most plausible explanations for supernovae.

Thermal Runaway

A white dwarf star may accumulate sufficient material from a stellar companion to raise its core temperature enough to ignite carbon fusion, at which point it undergoes runaway nuclear fusion, completely disrupting it. There are three avenues by which this detonation is theorized to happen: stable accretion of material from a companion, the collision of two white dwarfs, or accretion that causes ignition in a shell that then ignites. The dominant mechanism by which Type Ia supernovae are produced remains unclear. Despite this uncertainty in how Type Ia supernovae are produced, Type Ia supernovae have very uniform properties, and are useful standard candles over intergalactic distances. Some calibrations are required to compensate for the gradual change in properties or different frequencies of abnormal luminosity supernovae at high red shift, and for small variations in brightness identified by light curve shape or spectrum.

The progenitor of a Type Ia supernova

Two normal stars are in a binary pair.

The more massive star becomes a giant.

...which spills gas onto the secondary star, causing it to expand and become engulfed.

The secondary, lighter star and the core of the giant star spiral toward within a common envelope.

The common envelope is ejected, while the separation between the core and the secondary star decreases.

The remaining core of the giant collapses and becomes a white dwarf.

The aging companion star starts swelling, spilling gas onto the white dwarf.

The white dwarf's mass increases until it reaches a critical mass and explodes...

...causing the companion star to be ejected away.

Formation of a type Ia supernova

Normal Type Ia

There are several means by which a supernova of this type can form, but they share a common underlying mechanism. If a carbon-oxygen white dwarf accreted enough matter to reach the Chandrasekhar limit of about 1.44 solar masses ($M_\odot$) (for a non-rotating star), it would no longer be able to support the bulk of its mass through electron degeneracy pressure and would begin to collapse. However, the current view is that this limit is not normally attained; increasing temperature and density inside the core ignite carbon fusion as the star approaches the limit (to within about 1%), before collapse is initiated.

Within a few seconds, a substantial fraction of the matter in the white dwarf undergoes nuclear fusion, releasing enough energy (1−704420000000000000♠2×10^{44} J) to unbind the star in a supernova explosion. An outwardly expanding shock wave is generated, with matter reaching velocities on the order of 5,000–20,000 km/s, or roughly 3% of the speed of light. There is also a significant increase in luminosity, reaching an absolute magnitude of −19.3 (or 5 billion times brighter than the Sun), with little variation.

The model for the formation of this category of supernova is a closed binary star system. The larger of the two stars is the first to evolve off the main sequence, and it expands to form a red giant. The two stars now share a common envelope, causing their mutual orbit to shrink. The giant star then sheds most of its envelope, losing mass until it can no longer continue nuclear fusion. At this point it becomes a white dwarf star, composed primarily of carbon and oxygen. Eventually the secondary star also evolves off the main sequence to form a red giant. Matter from the giant is accreted by the white dwarf, causing the latter to increase in mass. Despite widespread acceptance of the basic model, the exact details of initiation and of the heavy elements produced in the explosion are still unclear.

Type Ia supernovae follow a characteristic light curve—the graph of luminosity as a function of time—after the explosion. This luminosity is generated by the radioactive decay of nickel-56 through cobalt-56 to iron-56. The peak luminosity of the light curve is extremely consistent across normal Type Ia supernovae, having a maximum absolute magnitude of about −19.3. This allows them to be used as a secondary standard candle to measure the distance to their host galaxies.

Non-Standard Type Ia

Another model for the formation of a Type Ia explosion involves the merger of two white dwarf stars, with the combined mass momentarily exceeding the Chandrasekhar limit. There is much variation in this type of explosion, and in many cases there may be no supernova at all, but it is expected that they will have a broader and less luminous light curve than the more normal Type Ia explosions.

Abnormally bright Type Ia supernovae are expected when the white dwarf already has a mass higher than the Chandrasekhar limit, possibly enhanced further by asymmetry, but the ejected material will have less than normal kinetic energy.

There is no formal sub-classification for the non-standard Type Ia supernovae. It has been proposed that a group of sub-luminous supernovae that occur when helium accretes onto a white dwarf should be classified as type Iax. This type of supernova may not always completely destroy the white dwarf progenitor and could leave behind a zombie star.

One specific type of non-standard Type Ia supernova develops hydrogen, and other, emission lines and gives the appearance of mixture between a normal Type Ia and a Type IIn supernova. Examples are SN 2002ic and SN 2005gj. These supernova have been dubbed Type Ia/IIn, Type Ian, Type IIa and Type IIan.

Core Collapse

Very massive stars can undergo core collapse when nuclear fusion suddenly becomes unable to sustain the core against its own gravity; this is the cause of all types of supernova except type Ia. The collapse may cause violent expulsion of the outer layers of the star resulting in a supernova, or the release of gravitational potential energy may be insufficient and the star may collapse into a black hole or neutron star with little radiated energy.

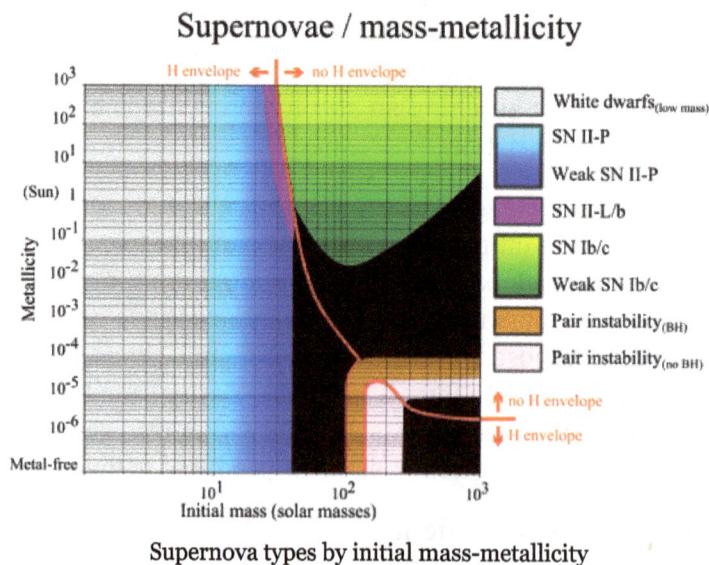

Supernova types by initial mass-metallicity

Core collapse can be caused by several different mechanisms: electron capture; exceeding the Chandrasekhar limit; pair-instability; or photodisintegration. When a massive star develops an iron core larger than the Chandrasekhar mass it will no longer be able to support itself by electron degeneracy pressure and will collapse further to a neutron star or black hole. Electron capture by magnesium in a degenerate O/Ne/Mg core causes gravitational collapse followed by explosive oxygen fusion, with very similar results. Electron-positron pair production in a large post-helium burning core removes thermodynamic support and causes initial collapse followed by runaway fusion, resulting in a pair-instability supernova. A sufficiently large and hot stellar core may generate gamma-rays energetic enough to initiate photodisintegration directly, which will cause a complete collapse of the core.

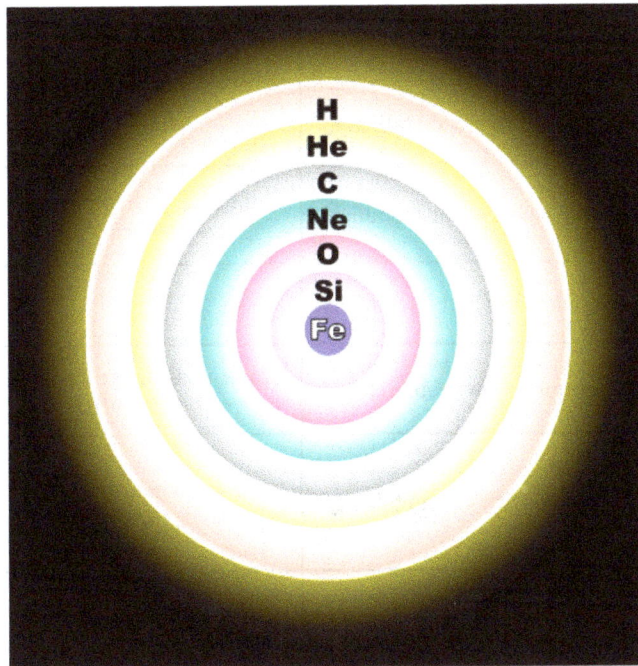

The onion-like layers of a massive, evolved star just prior to core collapse (Not to scale)

The table below lists the known reasons for core collapse in massive stars, the types of star that they occur in, their associated supernova type, and the remnant produced. The metallicity is the proportion of elements other than hydrogen or helium, as compared to the Sun. The initial mass is the mass of the star prior to the supernova event, given in multiples of the Sun's mass, although the mass at the time of the supernova may be much lower.

Type IIn supernovae are not listed in the table. They can potentially be produced by various types of core collapse in different progenitor stars, possibly even by type Ia white dwarf ignitions, although it seems that most will be from iron core collapse in luminous supergiants or hypergiants (including LBVs). The narrow spectral lines for which they are named occur because the supernova is expanding into a small dense cloud of circumstellar material. It appears that a significant proportion of supposed type IIn supernovae are actually supernova imposters, massive eruptions of LBV-like stars similar to the Great Eruption of Eta Carinae. In these events, material previously ejected from the star creates the narrow absorption lines and causes a shock wave through interaction with the newly ejected material.

Core collapse scenarios by mass and metallicity			
Cause of collapse	Progenitor star approximate initial mass	Supernova Type	Remnant

Electron capture in a degenerate O+Ne+Mg core	8–10	Faint II-P	Neutron star
Iron core collapse	10–25	Faint II-P	Neutron star
	25–40 with low or solar metallicity	Normal II-P	Black hole after fallback of material onto an initial neutron star
	25–40 with very high metallicity	II-L or II-b	Neutron star
	40–90 with low metallicity	None	Black hole
	≥40 with near-solar metallicity	Faint Ib/c, or hypernova with GRB	Black hole after fallback of material onto an initial neutron star
	≥40 with very high metallicity	Ib/c	Neutron star
	≥90 with low metallicity	None, possible gamma-ray burst (GRB)	Black hole
Pair instability	140–250 with low metallicity	II-P, sometimes a hypernova, possible GRB	No remnant
Photodisintegration	≥250 with low metallicity	None (or luminous supernova?), possible GRB	Massive black hole

Remnants of massive single stars

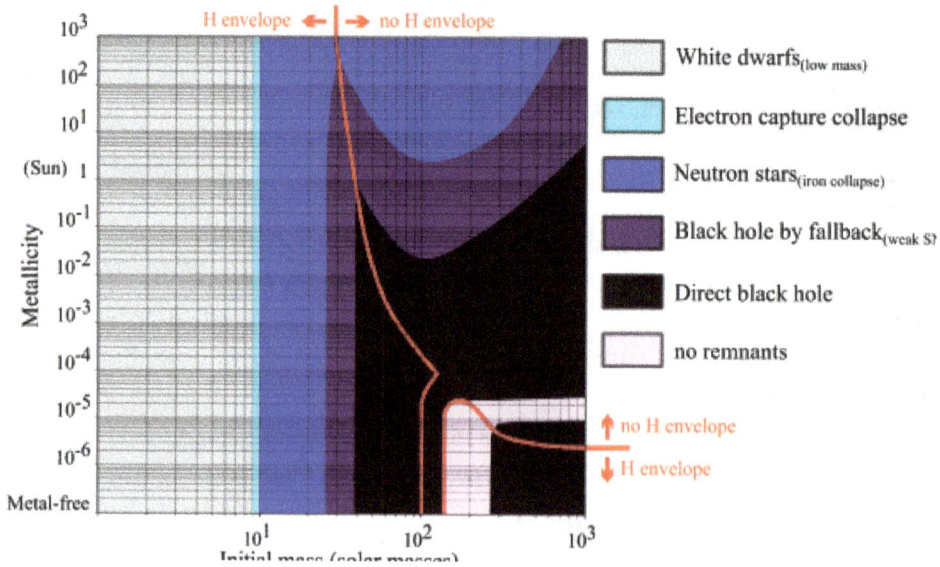

Remnants of single massive stars

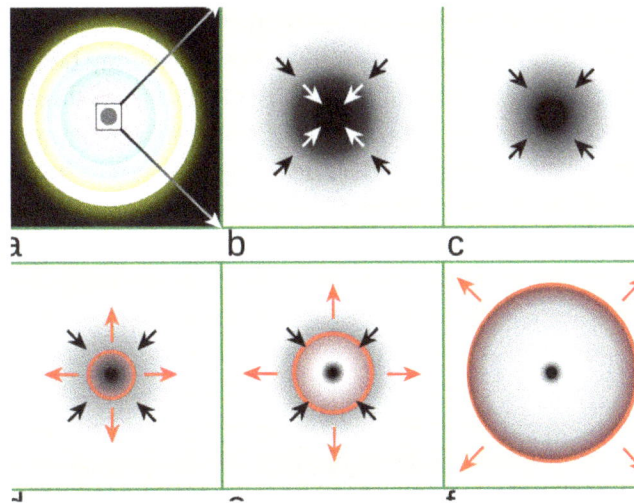

Within a massive, evolved star (a) the onion-layered shells of elements undergo fusion, forming an iron core (b) that reaches Chandrasekhar-mass and starts to collapse. The inner part of the core is compressed into neutrons (c), causing infalling material to bounce (d) and form an outward-propagating shock front (red). The shock starts to stall (e), but it is re-invigorated by a process that may include neutrino interaction. The surrounding material is blasted away (f), leaving only a degenerate remnant.

When a stellar core is no longer supported against gravity it collapses in on itself with velocities reaching 70,000 km/s (0.23c), resulting in a rapid increase in temperature and density. What follows next depends on the mass and structure of the collapsing core, with low mass degenerate cores forming neutron stars, higher mass degenerate cores mostly collapsing completely to black holes, and non-degenerate cores undergoing runaway fusion.

The initial collapse of degenerate cores is accelerated by beta decay, photodisintegration and electron capture, which causes a burst of electron neutrinos. As the density increases, neutrino emission is cut off as they become trapped in the core. The inner core eventually reaches typically 30 km diameter and a density comparable to that of an atomic nucleus, and neutron degeneracy pressure tries to halt the collapse. If the core mass is more than about $15\,M_\odot$ then neutron degeneracy is insufficient to stop the collapse and a black hole forms directly with no supernova explosion.

In lower mass cores the collapse is stopped and the newly formed neutron core has an initial temperature of about 100 billion kelvin, 6000 times the temperature of the sun's core. At this temperature, neutrino-antineutrino pairs of all flavors are efficiently formed by thermal emission. These thermal neutrinos are several times more abundant than the electron-capture neutrinos. About 10^{46} joules, approximately 10% of the star's rest mass, is converted into a ten-second burst of neutrinos which is the main output of the event. The suddenly halted core collapse rebounds and produces a shock wave that stalls within milliseconds in the outer core as energy is lost through the dissociation of heavy elements. A process that is not clearly understoodis necessary to allow the outer layers of the core to reabsorb around 10^{44} joules (1 foe) from the neutrino pulse, producing the visible explosion, although there are also other theories on how to power the explosion.

Some material from the outer envelope falls back onto the neutron star, and for cores beyond about $8\,M_\odot$ there is sufficient fallback to form a black hole. This fallback will reduce the kinetic energy of the explosion and the mass of expelled radioactive material, but in some situations it may also generate relativistic jets that result in a gamma-ray burst or an exceptionally luminous supernova.

Collapse of massive non-degenerate cores will ignite further fusion. When the core collapse is initiated by pair instability, oxygen fusion begins and the collapse may be halted. For core masses of 40–60 $M_\odot$, the collapse halts and the star remains intact, but core collapse will occur again when a larger core has formed. For cores of around 60–130 $M_\odot$, the fusion of oxygen and heavier elements is so energetic that the entire star is disrupted, causing a supernova. At the upper end of the mass range, the supernova is unusually luminous and extremely long-lived due to many solar masses of ejected ^{56}Ni. For even larger core masses, the core temperature becomes high enough to allow photodisintegration and the core collapses completely into a black hole.

Type II

The atypical subluminous type II SN 1997D

Stars with initial masses less than about eight times the sun never develop a core large enough to collapse and they eventually lose their atmospheres to become white dwarfs. Stars with at least 9 $M_\odot$ (possibly as much as 12 $M_\odot$) evolve in a complex fashion, progressively burning heavier elements at hotter temperatures in their cores. The star becomes layered like an onion, with the burning of more easily fused elements occurring in larger shells. Although popularly described as an onion with an iron core, the least massive supernova progenitors only have oxygen-neon(-magnesium) cores. These super AGB stars may form the majority of core collapse supernovae, although less luminous and so less commonly observed than those from more massive progenitors.

If core collapse occurs during a supergiant phase when the star still has a hydrogen envelope, the result is a Type II supernova. The rate of mass loss for luminous stars depends on the metallicity and luminosity. Extremely luminous stars at near solar metallicity will lose all their hydrogen before they reach core collapse and so will not form a type II supernova. At low metallicity, all stars will reach core collapse with a hydrogen envelope but sufficiently massive stars collapse directly to a black hole without producing a visible supernova.

Stars with an initial mass up to about 90 times the sun, or a little less at high metallicity, are expected to result in a type II-P supernova which is the most commonly observed type. At moderate to high metallicity, stars near the upper end of that mass range will have lost most of their hydrogen when core collapse occurs and the result will be a Type II-L supernova. At very low metallicity, stars of around 140–250 $M_\odot$ will reach core collapse by pair instability while they still have a hydrogen atmosphere and an oxygen core and the result will be a supernova with Type II characteristics but a very large mass of ejected Ni and high luminosity.

Type Ib and Ic

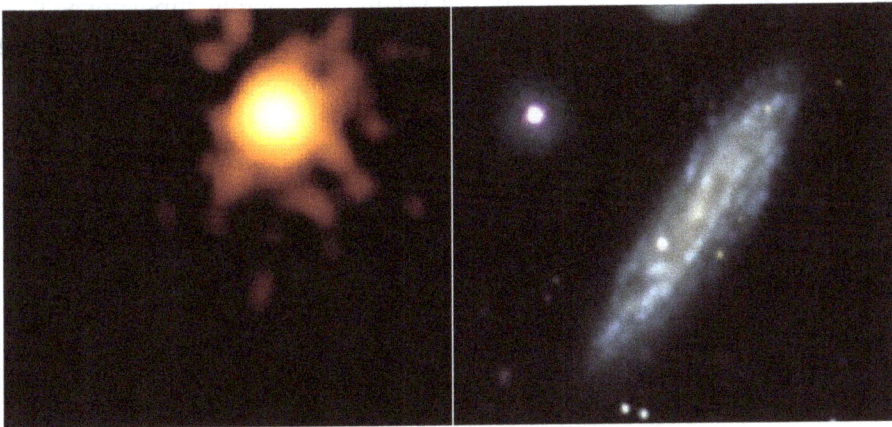

SN 2008D, a Type Ib supernova, shown in X-ray (left) and visible light (right) at the far upper end of the galaxy

These supernovae, like those of Type II, are massive stars that undergo core collapse. However the stars which become Types Ib and Ic supernovae have lost most of their outer (hydrogen) envelopes due to strong stellar winds or else from interaction with a companion. These stars are known as Wolf-Rayet stars, and they occur at moderate to high metallicity where continuum driven winds cause sufficiently high mass loss rates. Observations of type Ib/c supernova do not match the observed or expected occurrence of Wolf Rayet stars and alternate explanations for this type of core collapse supernova involve stars stripped of their hydrogen by binary interactions. Binary models provide a better match for the observed supernovae, with the proviso that no suitable binary helium stars have ever been observed. Since a supernova explosion can occur whenever the mass of the star at the time of core collapse is low enough not to cause complete fallback to a black hole, any massive star may result in a supernova if it loses enough mass before core collapse occurs.

Type Ib supernovae are the more common and result from Wolf-Rayet stars of type WC which still have helium in their atmospheres. For a narrow range of masses, stars evolve further before reaching core collapse to become WO stars with very little helium remaining and these are the progenitors of type Ic supernovae.

A few percent of the Type Ic supernovae are associated with gamma-ray bursts (GRB), though it is also believed that any hydrogen-stripped Type Ib or Ic supernova could produce a GRB, depending on the geometry of the explosion. The mechanism for producing this type of GRB is the jets produced by the magnetic field of the rapidly spinning magnetar formed at the collapsing core of the star. The jets would also transfer energy into the expanding outer shell of the explosion to produce a super-luminous supernova.

Ultra-stripped supernovae occur when the exploding star has been stripped (almost) all the way to the metal core, via mass transfer in a close binary. As a result, very little material is ejected from the exploding star (~0.1 M_{Sun}). In the most extreme cases, ultra-stripped supernovae can occur in naked metal cores, barely above the Chandrasekhar mass limit. SN 2005ek might be an observational example of an ultra-stripped supernova, giving rise to a relatively dim and fast decaying light curve. The nature of ultra-stripped supernovae can be both iron core-collapse and electron capture supernovae, depending on the mass of the collapsing core.

Failed

The core collapse of some massive stars may not result in a visible supernova. The main model for this is a sufficiently massive core that the explosion is insufficient to reverse the infall of the outer layers onto a black hole. These events are difficult to detect, but large surveys have detected possible candidates.

Light Curves

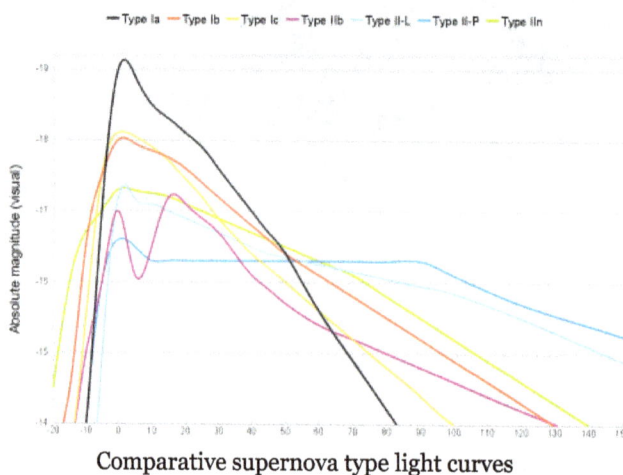

Comparative supernova type light curves

A historic puzzle concerned the source of energy that can maintain the optical supernova glow for months. Although the energy that disrupts each type of supernovae is delivered promptly, the light curves are mostly dominated by subsequent radioactive heating of the rapidly expanding ejecta. Some have considered rotational energy from the central pulsar. The ejecta gases would dim quickly without some energy input to keep it hot. The intensely radioactive nature of the ejecta gases, which is now known to be correct for most supernovae, was first calculated on sound nucleosynthesis grounds in the late 1960s. It was not until SN 1987A that direct observation of gamma-ray lines unambiguously identified the major radioactive nuclei.

It is now known by direct observation that much of the light curve (the graph of luminosity as a function of time) after the explosion of a Type II Supernova such as SN 1987A is provided its energy by those predicted radioactive decays. Although the luminous emission consists of optical photons, it is the radioactive power absorbed by the ejected gases that keeps the remnant hot enough to radiate light. The radioactive decay of through its daughters to produces gamma-ray photons , primarily of 847keV and 1238keV, that are absorbed and dominate the heating and thus the luminosity of the ejecta at intermediate times (several weeks) to late times (several months).

Energy for the peak of the light curve of SN1987A was provided by the decay of Ni to Co (half life 6 days) while energy for the later light curve in particular fit very closely with the 77.3 day half-life of Co decaying to Fe. Later measurements by space gamma-ray telescopes of the small fraction of the Co and Co gamma rays that escaped the SN 1987A remnant without absorption confirmed earlier predictions that those two radioactive nuclei were the power sources.

The visual light curves of the different supernova types all depend at late times on radioactive heating, but they vary in shape and amplitude on the underlying mechanisms of the explosion, the way that visible radiation is produced, the epoch of its observation, and the transparency of the ejected material. The light curves can be significantly different at other wavelengths. For example, at ultraviolet wavelengths there is an early extremely luminous peak lasting only a few hours corresponding to the breakout of the shock launched by the initial explosion, but that breakout is hardly detectable optically.

The light curves for type Ia are mostly very uniform, with a consistent maximum absolute magnitude and a relatively steep decline in luminosity. Their optical energy output is driven by radioactive decay of nickel-56 (half life 6 days), which then decays to radioactive cobalt-56 (half life 77 days). These radioisotopes from material ejected in the explosion excite surrounding material to incandescence. Studies of cosmology today rely on radioactivity providing the energy for the optical brightness of supernovae of Type Ia, which are the "standard candles" of cosmology but whose diagnostic 847keV and 1238keV gamma rays were first detected only in 2014. The initial phases of the light curve decline steeply as the effective size of the photosphere decreases and trapped electromagnetic radiation is depleted. The light curve continues to decline in the B band while it may show a small shoulder in the visual at about 40 days, but this is only a hint of a secondary maximum that occurs in the infra-red as certain ionised heavy elements recombine to produce infra-red radiation and the ejecta become transparent to it. The visual light curve continues to decline at a rate slightly greater than the decay rate of the radioactive cobalt (which has the longer half life and controls the later curve), because the ejected material becomes more diffuse and less able to convert the high energy radiation into visual radiation. After several months, the light curve changes its decline rate again as positron emission becomes dominant from the remaining cobalt-56, although this portion of the light curve has been little-studied.

Type Ib and Ic light curves are basically similar to type Ia although with a lower average peak luminosity. The visual light output is again due to radioactive decay being converted into visual radiation, but there is a much lower mass of nickel-56 produced in these types of explosion. The peak luminosity varies considerably and there are even occasional type Ib/c supernovae orders of magnitude more and less luminous than the norm. The most luminous type Ic supernovae are referred to as hypernovae and tend to have broadened light curves in addition to the increased peak luminosity. The source of the extra energy is thought to be relativistic jets driven by the formation of a rotating black hole, which also produce gamma-ray bursts.

The light curves for type II supernovae are characterised by a much slower decline than type I, on the order of 0.05 magnitudes per day, excluding the plateau phase. The visual light output is dominated by kinetic energy rather than radioactive decay for several months, due primarily to the existence of hydrogen in the ejecta from the atmosphere of the supergiant progenitor star. In the initial explosion this hydrogen becomes heated and ionised. The majority of type II supernovae show a prolonged plateau in their light curves as this hydrogen recombines, emitting visible light and becoming more transparent. This is then followed by a declining light curve driven by radioactive decay although slower than in type I supernovae, due to the efficiency of conversion into light by all the hydrogen.

In type II-L the plateau is absent because the progenitor had relatively little hydrogen left in its atmosphere, sufficient to appear in the spectrum but insufficient to produce a noticeable plateau in the light output. In type IIb supernovae the hydrogen atmosphere of the progenitor is so depleted (thought to be due to tidal stripping by a companion star) that the light curve is closer to a type I supernova and the hydrogen even disappears from the spectrum after several weeks.

Type IIn supernovae are characterised by additional narrow spectral lines produced in a dense shell of circumstellar material. Their light curves are generally very broad and extended, occasionally also extremely luminous and referred to as a hypernova. These light curves are produced by the highly efficient conversion of kinetic energy of the ejecta into electromagnetic radiation by interaction with the dense shell of material. This only occurs when the material is sufficiently dense and compact, indicating that it has been produced by the progenitor star itself only shortly before the supernova occurs.

Large numbers of supernovae have been catalogued and classified to provide distance candles and test models. Average characteristics vary somewhat with distance and type of host galaxy, but can broadly be specified for each supernova type.

Physical properties of supernovae by type				
Type[a]	Average peak absolute magnitude[b]	Approximate energy (foe)[c]	Days to peak luminosity	Days from peak to 10% luminosity
Ia	−19	1	approx. 19	around 60
Ib/c (faint)	around −15	0.1	15–25	unknown
Ib	around −17	1	15–25	40–100
Ic	around −16	1	15–25	40–100
Ic (bright)	to −22	above 5	roughly 25	roughly 100
II-b	around −17	1	around 20	around 100
II-L	around −17	1	around 13	around 150
II-P (faint)	around −14	0.1	roughly 15	unknown
II-P	around −16	1	around 15	Plateau then around 50
IIn[d]	around −17	1	12–30 or more	50–150
IIn (bright)	to −22	above 5	above 50	above 100

Asymmetry

The pulsar in the Crab nebula is travelling at 375 km/s relative to the nebula.

A long-standing puzzle surrounding Type II supernovae is why the compact object remaining after the explosion is given a large velocity away from the epicentre; pulsars, and thus neutron stars, are observed to have high velocities, and black holes presumably do as well, although they are far harder to observe in isolation. The initial impetus can be substantial, propelling an object of more than a solar mass at a velocity of 500 km/s or greater. This indicates an asymmetry in the explosion, but the mechanism by which momentum is transferred to the compact object remainsa puzzle. Proposed explanations for this kick include convection in the collapsing star and jet production during neutron star formation.

One possible explanation for the asymmetry in the explosion is large-scale convection above the core. The convection can create variations in the local abundances of elements, resulting in uneven nuclear burning during the collapse, bounce and resulting explosion.

Another possible explanation is that accretion of gas onto the central neutron star can create a disk that drives highly directional jets, propelling matter at a high velocity out of the star, and driving transverse shocks that completely disrupt the star. These jets might play a crucial role in the resulting supernova explosion. (A similar model is now favored for explaining long gamma-ray bursts.)

Initial asymmetries have also been confirmed in Type Ia supernova explosions through observation. This result may mean that the initial luminosity of this type of supernova depends on the viewing angle. However, the explosion becomes more symmetrical with the passage of time. Early asymmetries are detectable by measuring the polarization of the emitted light.

Energy Output

Although we are used to thinking of supernovae primarily as luminous visible events, the electromagnetic radiation they release is almost a minor side-effect of the explosion. Particularly in the case of core collapse supernovae, the emitted electromagnetic radiation is a tiny fraction of the total energy released during the event.

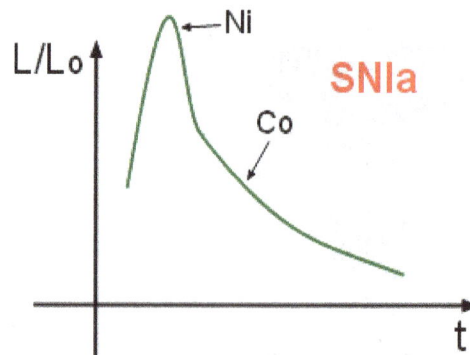

The radioactive decays of nickel-56 and cobalt-56 that produce a supernova visible light curve

There is a fundamental difference between the balance of energy production in the different types of supernova. In type Ia white dwarf detonations, most of the energy is directed into heavy element synthesis and the kinetic energy of the ejecta. In core collapse supernovae, the vast majority of the energy is directed into neutrino emission, and while some of this apparently powers the following main explosion 99%+ of the neutrinos escape the star in the first few minutes following the start of the collapse.

Type Ia supernovae derive their energy from a runaway nuclear fusion of a carbon-oxygen white dwarf. The details of the energetics are still not fully understood, but the end result is the ejection of the entire mass of the original star at high kinetic energy. Around half a solar mass of that mass is Ni generated from silicon burning. Ni is radioactive and decays into Co by beta plus decay (with a half life of six days) and gamma rays. Co itself decays by the beta plus (an anti-electron) path with a half life of 77 days into stable Fe. These two processes are responsible for the electromagnetic radiation from type Ia supernovae. In combination with the changing transparency of the ejected material, they produce the rapidly declining light curve.

Core collapse supernovae are on average visually fainter than type Ia supernovae, but the total energy released is far higher. In the case of supernovae, the energy of gravitational potential energy is converted into kinetic energy that compresses and collapses the core, initially producing electron neutrinos from disintegrating nucleons, followed by all flavours of thermal neutrinos from the super-heated neutron star core. Around 1% of these neutrinos are thought to deposit sufficient energy into the outer layers of the star to drive the resulting explosion, but again the details cannot be reproduced exactly in current models. Kinetic energies and nickel yields are somewhat lower than type Ia supernovae, hence the lower visual luminosity of supernovae, but energy from the de-ionisation of the many solar masses of remaining hydrogen can contribute to a much slower decline in luminosity and produce the plateau phase seen in the majority of core collapse supernovae.

Energetics of supernovae					
Supernova	Approximate total energy 10^{44} joules (foe)[c]	Ejected Ni (solar masses)	Neutrino energy (foe)	Kinetic energy (foe)	Electromagnetic radiation (foe)
Type Ia	1.5	0.4 − 0.8	0.1	1.3 − 1.4	~0.01

Core collapse	100	$(0.01) - 1$	100	1	$0.001 - 0.01$
Hypernova	100	~1	1–100	1–100	~0.1
Pair instability	5–100	$0.5 - 50$	low?	1–100	$0.01 - 0.1$

In some core collapse supernovae, fallback onto a black hole drives relativistic jets which may produce a brief energetic and directional burst of gamma rays and also transfers substantial further energy into the ejected material. This is one scenario for producing high luminosity supernovae and is thought to be the cause of type Ic hypernovae and long duration gamma-ray bursts. If the relativistic jets are too brief and fail to penetrate the stellar envelope then a low luminosity gamma-ray burst may be produced and the supernova may be sub-luminous.

When a supernova occurs inside a small dense cloud of circumstellar material, it will produce a shock wave that can efficiently convert a high fraction of the kinetic energy into electromagnetic radiation. Even though the initial explosion energy was entirely normal the resulting supernova will have high luminosity and extended duration since it does not rely on exponential radioactive decay. This type of event may cause type IIn hypernovae.

Although pair-instability supernovae are core collapse supernovae with spectra and light curves similar to type II-P, the nature of that explosion following core collapse is more like that of a giant type Ia with runaway fusion of carbon, oxygen, and silicon. The total energy released by the highest mass events is comparable to other core collapse supernovae but neutrino production is thought to be very low, hence the kinetic and electromagnetic energy released is very high. The cores of these stars are much larger than any white dwarf and the amount of radioactive nickel and other heavy elements ejected from their cores can be orders of magnitude higher, with consequently high visual luminosity.

Progenitor

The supernova classification type is closely tied to the type of star at the time of the explosion. The occurrence of each type of supernova depends dramatically on the metallicity and hence the age of the host galaxy.

Shown in this sped-up artist's impression, is a collection of distant galaxies, the occasional supernova can be seen. Each of these exploding stars briefly rivals the brightness of its host galaxy.

Type Ia supernovae are produced from white dwarf stars in binary systems and occur in all galaxy types. Core collapse supernovae are only found in galaxies undergoing current or very recent star formation, since they result from short-lived massive stars. They are most commonly found in type Sc spirals, but also in the arms of other spiral galaxies and in irregular galaxies, especially starburst galaxies.

Type Ib/c and II-L, and possibly most type IIn, supernovae are only thought to be produced from stars having near-solar metallicity levels that result in high mass loss from massive stars, hence they are less common in older more distant galaxies. The table shows the expected progenitor for the main types of core collapse supernova, and the approximate proportions of each in the local neighbourhood.

Fraction of core collapse supernovae types by progenitor		
Type	Progenitor star	Fraction
Ib	WC Wolf-Rayet	9%
Ic	WO Wolf-Rayet	17%
II-P	Supergiant	55%
II-L	Supergiant with a depleted hydrogen shell	3.0%
IIn	Supergiant in a dense cloud of expelled material (such as LBV)	2.4%
IIb	Supergiant with highly depleted hydrogen (stripped by companion?)	12%
IIpec	Blue supergiant?	1.0%

There are a number of difficulties reconciling modelled and observed stellar evolution leading up to core collapse supernovae. Red supergiants are the expected progenitors for the vast majority of core collapse supernovae, and these have been observed but only at relatively low masses and luminosities, below about $18\,M_\odot$ and $100{,}000\,L_\odot$ respectively. Most progenitors of type II supernovae are not detected and must be considerably fainter, and presumably less massive. It is now proposed that higher mass red supergiants do not explode as supernovae, but instead evolve back towards hotter temperatures. Several progenitors of type IIb supernovae have been confirmed, and these were K and G supergiants, plus one A supergiant. Yellow hypergiants or LBVs are proposed progenitors for type IIb supernovae, and almost all type IIb supernovae near enough to observe have shown such progenitors.

Until just a few decades ago, hot supergiants were not considered likely to explode, but observations have shown otherwise. Blue supergiants form an unexpectedly high proportion of confirmed supernova progenitors, partly due to their high luminosity and easy detection, while not a single Wolf-Rayet progenitor has yet been clearly identified. Models have had difficulty showing how blue supergiants lose enough mass to reach supernova without progressing to a different evolutionary stage. One study has shown a possible route for low-luminosity post-red supergiant luminous blue variables to collapse, most likely as a type IIn supernova.

The expected progenitors of type Ib supernovae, luminous WC stars, are not observed at all. Instead WC stars are found at lower luminosities, apparently post-red supergiant stars. WO stars are extremely rare and visually relatively faint, so it is difficult to say whether such progenitors are missing or just yet to be observed. Very luminous progenitors, despite numerous supernovae being observed near enough that such progenitors would have been clearly imaged. Several examples of hot luminous progenitors of type IIn supernovae have been detected: SN 2005gy and SN 2010jl were both apparently massive luminous stars, but are very distant; and SN 2009ip had a highly luminous progenitor likely to have been an LBV, but is a peculiar supernova whose exact nature is disupted.

Interstellar Impact

Source of Heavy Elements

Supernovae are the major source of elements heavier than oxygen. These elements are produced by nuclear fusion for nuclei up to S, by silicon photodisintegration rearrangement and quasiequilibrium during silicon burning for nuclei between Ar and Ni, and by rapid captures of neutrons during the supernova explosion for elements heavier than iron. Nucleosynthesis during silicon burning yields nuclei roughly 1000-100,000 times more abundant than the r-process isotopes heavier than iron. Supernovae are the most likely, although not undisputed, candidate sites for the r-process, which is the rapid capture of neutrons that occurs at high temperature and high density of neutrons. Those reactions produce highly unstable nuclei that are rich in neutrons and that rapidly beta decay into more stable forms. The r-process produces about half of all the heavier isotopes of the elements beyond iron, including plutonium and uranium. The only other major competing process for producing elements heavier than iron is the s-process in large, old, red-giant AGB stars, which produces these elements slowly over longer epochs and which cannot produce elements heavier than lead.

Role in Stellar Evolution

The remnant of a supernova explosion consists of a compact object and a rapidly expanding shock wave of material. This cloud of material sweeps up the surrounding interstellar medium during a free expansion phase, which can last for up to two centuries. The wave then gradually undergoes a period of adiabatic expansion, and will slowly cool and mix with the surrounding interstellar medium over a period of about 10,000 years.

Supernova remnant N 63A lies within a clumpy region of gas and dust in the Large Magellanic Cloud.

The Big Bang produced hydrogen, helium, and traces of lithium, while all heavier elements are synthesized in stars and supernovae. Supernovae tend to enrich the surrounding interstellar medium with *metals*—elements other than hydrogen and helium.

These injected elements ultimately enrich the molecular clouds that are the sites of star formation. Thus, each stellar generation has a slightly different composition, going from an almost pure mixture of hydrogen and helium to a more metal-rich composition. Supernovae are the dominant mechanism for distributing these heavier elements, which are formed in a star during its period of nuclear fusion. The different abundances of elements in the material that forms a star have important influences on the star's life, and may decisively influence the possibility of having planets orbiting it.

The kinetic energy of an expanding supernova remnant can trigger star formation due to compression of nearby, dense molecular clouds in space. The increase in turbulent pressure can also prevent star formation if the cloud is unable to lose the excess energy.

Evidence from daughter products of short-lived radioactive isotopes shows that a nearby supernova helped determine the composition of the Solar System 4.5 billion years ago, and may even have triggered the formation of this system. Supernova production of heavy elements over astronomic periods of time ultimately made the chemistry of life on Earth possible.

Effect on Earth

A near-Earth supernova is a supernova close enough to the Earth to have noticeable effects on its biosphere. Depending upon the type and energy of the supernova, it could be as far as 3000 light-years away. Gamma rays from a supernova would induce a chemical reaction in the upper atmosphere converting molecular nitrogen into nitrogen oxides, depleting the ozone layer enough to expose the surface to harmful solar radiation. This has been proposed as the cause of the Ordovician–Silurian extinction, which resulted in the death of nearly 60% of the oceanic life on Earth. In 1996 it was theorized that traces of past supernovae might be detectable on Earth in the form of metal isotope signatures in rock strata. Iron-60 enrichment was later reported in deep-sea rock of the Pacific Ocean. In 2009, elevated levels of nitrate ions were found in Antarctic ice, which coincided with the 1006 and 1054 supernovae. Gamma rays from these supernovae could have boosted levels of nitrogen oxides, which became trapped in the ice.

Type Ia supernovae are thought to be potentially the most dangerous if they occur close enough to the Earth. Because these supernovae arise from dim, common white dwarf stars in binary systems, it is likely that a supernova that can affect the Earth will occur unpredictably and in a star system that is not well studied. The closest known candidate is IK Pegasi. Recent estimates predict that a Type II supernova would have to be closer than eight parsecs (26 light-years) to destroy half of the Earth's ozone layer, and there are no such candidates closer than about 500 light years.

Milky Way Candidates

The next supernova in the Milky Way will likely be detectable even if it occurs on the far side of the galaxy. It is likely to be produced by the collapse of an unremarkable red supergiant and it is very probable that it will already have been catalogued in infrared surveys such as 2MASS. There is a smaller chance that the next core collapse supernova will be produced by a different type of massive star such as a yellow hypergiant, luminous blue variable, or Wolf-Rayet. The chances of the next supernova being a type Ia produced by a white dwarf are calculated to be about a third of those for a core collapse supernova. Again it should be observable wherever it occurs, but it is less likely that the progenitor will ever have been observed prior to the explosion. It isn't even known exactly what a type Ia progenitor system looks like, and difficult to detect them beyond a few parsecs. The total supernova rate in our galaxy is estimated to be about 4.6 per century, or one every 22 years, although we haven't actually observed one for several centuries.

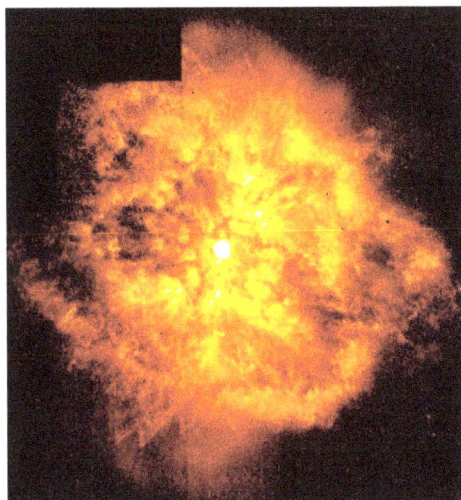

The nebula around Wolf–Rayet star WR124, which is located at a distance of about 21,000 light years.

Statistically, the next supernova is likely to be produced from an otherwise unremarkable red supergiant, but it is difficult to identify which of those supergiants are in the final stages of heavy element fusion in their cores and which have millions of years left. The most massive red supergiants are expected to shed their atmospheres and evolve to Wolf-Rayet stars before their cores collapse. All Wolf-Rayet stars are expected to end their lives from the Wolf-Rayet phase within a million years or so, but again it is difficult to identify those that are closest to core collapse. One class that is expected to have no more than a few thousand years before exploding are the WO Wolf-Rayet stars, which are known to have exhausted their core helium. Only eight of them are known, and only four of those are in the Milky Way.

A number of close or well known stars have been identified as possible core collapse supernova candidates: the red supergiants Antares and Betelgeuze; the yellow hypergiant Rho Cassiopeiae; the luminous blue variable Eta Carinae that has already produced a supernova imposter explosion; and the brightest component, a Wolf–Rayet star in the Regor or Gamma Velorum system, Others have gained notoriety as possible, although not very likely, progenitors for a gamma-ray burst; for example WR 104.

Identification of candidates for a type Ia supernova explosion is much more speculative. Any binary with an accreting white dwarf might produce a supernova although the exact mechanism and timescale is still debated. These systems are faint and difficult to identify, but the novae and recurrent novae are such systems that conveniently advertise themselves. One examples is U Scorpii, The nearest known type Ia supernova candidate is IK Pegasi (HR 8210), located at a distance of 150 light-years, but observations suggest it will be several million years before the white dwarf can accrete the critical mass required to become a Type Ia supernova.

Protostar

A protostar is a very young star that is still gathering mass from its parent molecular cloud. The protostellar phase is the earliest one in the process of stellar evolution. For a one so-lar-mass star it lasts about 1,000,000 years. The phase begins when a molecular cloud first collapses under the force of self-gravity. It ends when the protostar blows back the infalling gas and is revealed as an optically visible pre-main-sequence star, which later contracts to become a main sequence star.

History

The modern picture of protostars, summarized above, was first suggested by Chushiro Hayashi. In the first models, the size of protostars was greatly overestimated. Subsequent numerical calculations clarified the issue, and showed that protostars are only modestly larger than main-sequence stars of the same mass. This basic theoretical result has been confirmed by observations, which find that the largest pre-main-sequence stars are also of modest size.

Protostellar Evolution

Star formation begins in relatively small molecular clouds called dense cores. Each dense core is initially in balance between self-gravity, which tends to compress the object, and both gas pressure and magnetic pressure, which tend to inflate it. As the dense core accrues mass from its larger, surrounding cloud, self-gravity begins to overwhelm pressure, and collapse begins. Theoretical modeling of an idealized spherical cloud initially supported only by gas pressure indicates that the collapse process spreads from the inside toward the outside. Spectroscopic observations of dense cores that do not yet contain stars indicate that contraction indeed occurs. So far, however, the predicted outward spread of the collapse region has not been observed.

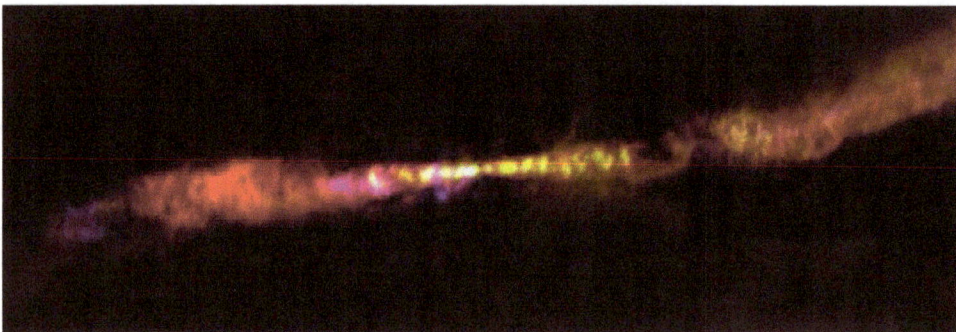

Infant star CARMA-7 and its jets are located approximately 1400 light-years from Earth within the Serpens South star cluster.

The gas that collapses toward the center of the dense core first builds up a low-mass protostar, and then a protoplanetary disk orbiting the object. As the collapse continues, an increasing amount of gas impacts the disk rather than the star, a consequence of angular momentum conservation. Exactly how material in the disk spirals inward onto the protostar is not yet understood, despite a great deal of theoretical effort. This problem is illustrative of the larger issue of accretion disk theory, which plays a role in much of astrophysics.

HBC 1 is a young pre-main-sequence star.

Regardless of the details, the outer surface of a protostar consists at least partially of shocked gas that has fallen from the inner edge of the disk. The surface is thus very different from the relatively quiescent photosphere of a pre-main sequence or main-sequence star. Within its deep interior, the protostar has lower temperature than an ordinary star. At its center, hydrogen is not yet undergoing nuclear fusion. Theory predicts, however, that the hydrogen isotope deuterium is undergoing fusion, creating helium-3. The heat from this fusion reaction tends to inflate the protostar, and thereby helps determine the size of the youngest observed pre-main-sequence stars.

The energy generated from ordinary stars comes from the nuclear fusion occurring at their centers. Protostars also generate energy, but it comes from the radiation liberated at the shocks on its surface and on the surface of its surrounding disk. The radiation thus created must traverse the interstellar dust in the surrounding dense core. The dust absorbs all impinging photons and reradiates them at longer wavelengths. Consequently, a protostar is not detectable at optical wavelengths, and cannot be placed in the Hertzsprung-Russell diagram, unlike the more evolved pre-main-sequence stars.

The actual radiation emanating from a protostar is predicted to be in the infrared and millimeter regimes. Point-like sources of such long-wavelength radiation are commonly seen in regions that are obscured by molecular clouds. It is commonly believed that those conventionally labeled as Class 0 or Class I sources are protostars. However, there is still no definitive evidence for this identification.

Chemical Composition

The Sun as a protostar had the same composition as today, which is 71.1% hydrogen, 27.4% helium, and 1.5% heavier elements, by mass.

Observed Classes of Young Stars

For details of observational classification, see Table below:

Class	peak emission	duration (Years)
O	submillimeter	10^4
I	far-infrared	10^5
II	near-infrared	10^6
III	visible	10^7

Red Dwarf

Proxima Centauri, the closest star to the Sun at 4.2 ly, is a red dwarf

A red dwarf is a small and relatively cool star on the main sequence, of either K or M spectral type. Red dwarfs range in mass from a low of 0.075 solar masses ($M_\odot$) to about 0.50 $M_\odot$ and have a surface temperature of less than 4,000 K.

Red dwarfs are by far the most common type of star in the Milky Way, at least in the neighborhood of the Sun, but because of their low luminosity, individual red dwarfs cannot be easily observed. From Earth, not one is visible to the naked eye. Proxima Centauri, the nearest star to the Sun, is a red dwarf (Type M5, apparent magnitude 11.05), as are twenty of the next thirty nearest stars. According to some estimates, red dwarfs make up three-quarters of the stars in the Milky Way.

Stellar models indicate that red dwarfs less than 0.35 $M_\odot$ are fully convective. Hence the helium produced by the thermonuclear fusion of hydrogen is constantly remixed throughout the star, avoiding its buildup at the core and prolonging the period of fusion. Red dwarfs therefore develop very slowly, maintaining a constant luminosity and spectral type for trillions of years, until their fuel is depleted. Because of the comparatively short age of the universe, no red dwarfs exist at advanced stages of evolution.

Definition

The term red dwarf when used to refer to a star does not have a strict definition. One of the earliest uses of the term was in 1915, used simply to contrast "red" dwarf stars from hotter "blue" dwarf stars. It became established use, although the definition remained vague. In terms of which spectral types qualify as red dwarfs, different researchers picked different limits, for example K8–M5 or "later than K5". *Dwarf M star*, abbreviated dM, was also used, but sometimes it also included stars of spectral type K.

In modern usage, the definition of a *red dwarf* still varies. When explicitly defined, it typically includes late K- and early to mid-M-class stars, but in many cases it is restricted just to M-class stars. In some cases all K stars are included as red dwarfs, and occasionally even earlier stars.

The coolest true main-sequence stars are thought to have spectral types around L2 or L3, but many objects cooler than about M6 or M7 are brown dwarfs, insufficiently massive to sustain hydrogen-1 fusion.

Description and Characteristics

Red dwarfs are very-low-mass stars. As a result, they have relatively low pressures, a low fusion rate, and hence, a low temperature. The energy generated is the product of nuclear fusion of hydrogen into helium by way of the proton–proton (PP) chain mechanism. Hence, these stars emit little light, sometimes as little as $\frac{1}{10,000}$ that of the Sun. Even the largest red dwarfs (for example HD 179930, HIP 12961 and Lacaille 8760) have only about 10% of the Sun's luminosity. In general, red dwarfs less than 0.35 $M_\odot$ transport energy from the core to the surface by convection. Convection occurs because of opacity of the interior, which has a high density compared to the temperature. As a result, energy transfer by radiation is decreased, and instead convection is the main form of energy transport to the surface of the star. Above this mass, a red dwarf will have a region around its core where convection does not occur.

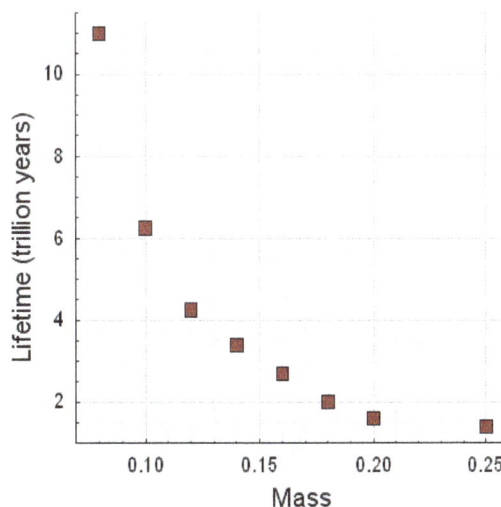

The predicted main-sequence lifetime of a red dwarf plotted against its mass relative to the Sun.

Because low-mass red dwarfs are fully convective, helium does not accumulate at the core, and compared to larger stars such as the Sun, they can burn a larger proportion of their hydrogen before leaving the main sequence. As a result, red dwarfs have estimated lifespans far longer than the present age of the universe, and stars less than 0.8 $M_\odot$ have not had time to leave the main sequence. The lower the mass of a red dwarf, the longer the lifespan. It is believed that the lifespan of these stars exceeds the expected 10 billion year lifespan of our Sun by the third or fourth power of the ratio of the solar mass to their masses; thus a 0.1 $M_\odot$ red dwarf may continue burning for 10 trillion years. As the proportion of hydrogen in a red dwarf is consumed, the rate of fusion declines and the core starts to contract. The gravitational energy released by this size reduction is converted into heat, which is carried throughout the star by convection.

Typical characteristics				
Stellar class	Mass ($M_\odot$)	Radius ($R_\odot$)	Luminosity ($L_\odot$)	Teff (K)
M0V	60%	62%	7.2%	3,800
M1V	49%	49%	3.5%	3,600
M2V	44%	44%	2.3%	3,400
M3V	36%	39%	1.5%	3,250
M4V	20%	26%	0.55%	3,100
M5V	14%	20%	0.22%	2,800
M6V	10%	15%	0.09%	2,600
M7V	9%	12%	0.05%	2,500
M8V	8%	11%	0.03%	2,400
M9V	7.5%	8%	0.015%	2,300

According to computer simulations, the minimum mass a red dwarf must have in order to eventually evolve into a red giant is 0.25 $M_\odot$; less massive objects, as they age, would increase their surface temperatures and luminosities becoming blue dwarfs and finally white dwarfs.

The less massive the star, the longer this evolutionary process takes. It has been calculated that a 0.16 $M_\odot$ red dwarf (approximately the mass of the nearby Barnard's Star) would stay on the main sequence for 2.5 trillion years, followed by five billion years as a blue dwarf, during which the star would have one third of the Sun's luminosity ($L_\odot$) and a surface temperature of 6,500□8,500 Kelvin.

Scattered stars in Sagittarius. Red dwarfs and are thought to be the most common type of star within the Milky Way.

The fact that red dwarfs and other low-mass stars still remain on the main sequence when more massive stars have moved off the main sequence allows the age of star clusters to be estimated by finding the mass at which the stars move off the main sequence. This provides a lower limit to the age of the Universe and also allows formation timescales to be placed upon the structures within the Milky Way, such as the Galactic halo and Galactic disk.

All observed red dwarfs contain "metals", which in astronomy are elements heavier than hydrogen and helium. The Big Bang model predicts that the first generation of stars should have only hydrogen, helium, and trace amounts of lithium, and hence would be of low metallicity. With their extreme lifespans, any red dwarfs that were a part of that first generation (population III stars) should still exist today. Low metallicity red dwarfs, however, are rare. There are several explanations for the missing population of metal-poor red dwarfs. The preferred explanation is that, without heavy elements, only large stars can form. Large stars rapidly burn out and explode as supernova, spewing heavy elements that then allow higher metallicity stars population II stars, including red dwarfs to form. Alternative explanations of the scarcity of metal-poor red dwarfs, such as their dimness and scarcity, are considered less likely because they appear to conflict with stellar-evolution models.

Planets

Many red dwarfs are orbited by exoplanets, but large Jupiter-sized planets are comparatively rare. Doppler surveys of a wide variety of stars indicate about 1 in 6 stars with twice the mass of the Sun are orbited by one or more Jupiter-sized planets, vs. 1 in 16 for Sun-like stars and only 1 in 50 for red dwarfs. On the other hand, microlensing surveys indicate that long-orbital-period Neptune-mass planets are found around one in three red dwarfs. Observations with HARPS further indicate 40% of red dwarfs have a "super-Earth" class planet orbiting in the habitable zone where liquid water can exist on the surface.

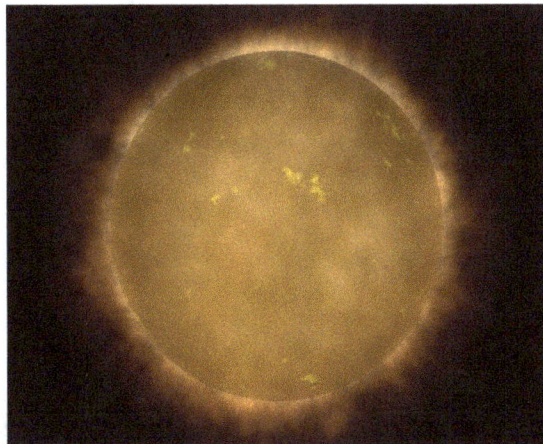

Artist's conception of a red dwarf, the most common type of star in the Sun's stellar neighborhood, and in the universe. Although termed a red dwarf, the surface temperature of this star would give it an orange hue when viewed from close proximity

At least four and possibly up to six exoplanets were discovered orbiting the red dwarf Gliese 581 between 2005–2010. One planet has about the mass of Neptune, or 16 Earth masses ($M_\square$). It orbits just 6 million kilometers (0.04 AU) from its star, and is estimated to have a surface temperature of 150 °C, despite the dimness of its star. In 2006, an even smaller exoplanet (only 5.5 $M_\oplus$) was found orbiting the red dwarf OGLE-2005-BLG-390L; it lies 390 million km (2.6 AU) from the star and its surface temperature is −220 °C (56 K).

In 2007, a new, potentially habitable exoplanet, Gliese 581 c, was found, orbiting Gliese 581. If the minimum mass estimated by its discoverers (a team led by Stephane Udry), of 5.36 $M_\oplus$, is correct, it is the smallest exoplanet orbiting a main-sequence star discovered to date and since then Gliese 581 d, which is also potentially habitable, was discovered. (There are smaller planets known around a neutron star, PSR B1257+12.) The discoverers estimate its radius to be 1.5 times that of Earth ($R_\oplus$).

Gliese 581 c and d are within the habitable zone of the host star, and are two of the most likely candidates for habitability of any exoplanets discovered so far. Gliese 581 g, detected September 2010, has a near-circular orbit in the middle of the star's habitable zone. However, the planet's existence is contested.

Habitability

Planetary habitability of red dwarf systems is subject to some debate. In spite of their great numbers and long lifespans, there are several factors which may make life difficult on planets around a red dwarf. First, planets in the habitable zone of a red dwarf would be so close to the parent star that they would likely be tidally locked. This would mean that one side would be in perpetual daylight and the other in eternal night. This could create enormous temperature variations from one side of the planet to the other. Such conditions would appear to make it difficult for forms of life similar to those on Earth to evolve. And it appears there is a great problem with the atmosphere of such tidally locked planets: the perpetual night zone would be cold enough to freeze the main gases of their atmospheres, leaving the daylight zone nude and dry. On the other hand, recent theories propose that either a thick atmosphere or planetary ocean could potentially circulate heat around such a planet. Alternatively, a moon in orbit around a gas giant may be habitable. It would circumvent the tidal lock problem with its star by becoming tidally locked to its planet. In this manner there would be a day/night cycle as the moon orbited its primary, and hence a more uniform distribution of heat.

An artist's impression of a planet with two exomoons orbiting in the habitable zone of a red dwarf.

An additional difficulty is that red dwarfs radiate most of their electromagnetic energy as infrared light, whereas plants on Earth capture most of their energy from the visible spectrum. Red dwarfs emit almost no ultraviolet light, which would be a problem, should this kind of light be required for life to exist. Variability in stellar energy output may also have negative impacts on the development of life. Red dwarfs are often covered by starspots, reducing stellar output by as much as 40%

for months at a time. At other times, some red dwarfs, called flare stars, can emit gigantic flares, doubling their brightness in minutes. This variability may also make it difficult for life to develop and persist near a red dwarf. It may be possible for a planet orbiting close to a red dwarf to keep its atmosphere even if the star flares. However, more-recent research suggests that these stars may be the source of constant high energy flares and very large magnetic fields, diminishing the possibility of life as we know it. Whether this is a peculiarity of the star under examination or a feature of the entire class remains to be determined.

Spectral Standard Stars

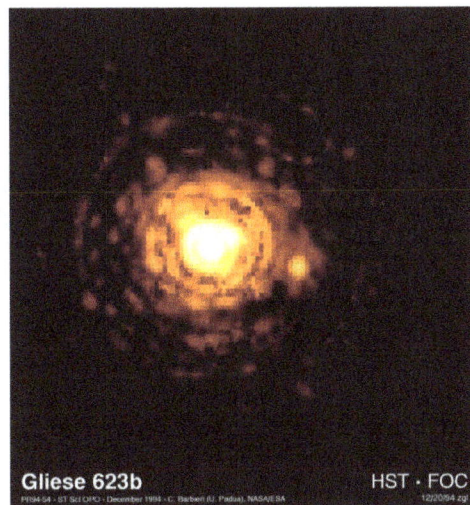

Gliese 623b is right of center

The spectral standards for M-type stars have changed slightly over the years, but settled down somewhat since the early 1990s. Part of this is due to the fact that even the nearest red dwarfs are fairly faint, and the study of mid- to late-M dwarfs has progressed only in the past few decades due to evolution of astronomical techniques, from photographic plates to charged-couple devices (CCDs) to infrared-sensitive arrays.

The revised Yerkes Atlas system (Johnson & Morgan 1953) listed only 2 M-type spectral standard stars: HD 147379 (M0 V) and HD 95735/Lalande 21185 (M2 V). While HD 147379 was not considered a standard by expert classifiers in later compendia of standards, Lalande 21185 is still a primary standard for M2 V. Robert Garrison does not list any "anchor" standards among the red dwarfs, but Lalande 21185 has survived as a M2 V standard through many compendia. The review on MK classification by Morgan & Keenan (1973) did not contain red dwarf standards. In the mid-1970s, red dwarf standard stars were published by Keenan & McNeil (1976) and Boeshaar (1976), but unfortunately there was little agreement among the standards. As later cooler stars were identified through the 1980s, it was clear that an overhaul of the red dwarf standards was needed. Building primarily upon the Boeshaar standards, a group at Steward Observatory (Kirkpatrick, Henry, & McCarthy 1991) filled in the spectral sequence from K5 V to M9 V. It is these M-type dwarf standard stars which have largely survived as the main standards to the modern day. There have been negligible changes in the red dwarf spectral sequence since 1991. Additional red dwarf standards were compiled by Henry et al. (2002), and D. Kirkpatrick has recently reviewed the classification of red dwarfs and standard stars in Gray & Corbally's 2009 monograph. The M-dwarf primary spectral standards are: GJ 270 (M0 V), GJ 229A (M1 V), Lalande 21185 (M2 V), Gliese 581 (M3 V), GJ 402 (M4 V), GJ 51 (M5 V), Wolf 359 (M6 V), Van Biesbroeck 8 (M7 V), VB 10 (M8 V), LHS 2924 (M9 V).

White Dwarf

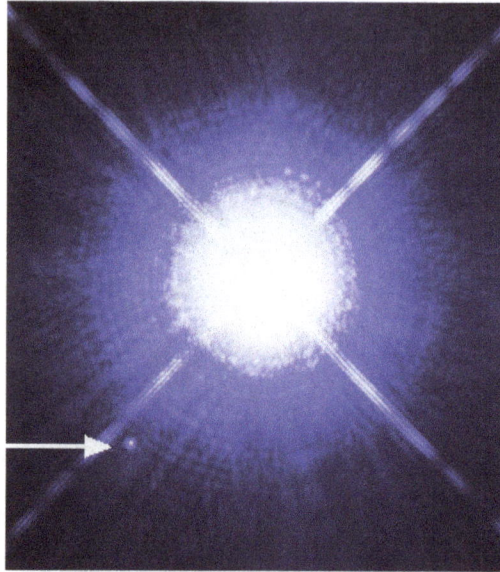

Image of Sirius A and Sirius B taken by the Hubble Space Telescope. Sirius B, which is a white dwarf, can be seen as a faint pinprick of light to the lower left of the much brighter Sirius A.

Artist's concept of white dwarf aging

A white dwarf, also called a degenerate dwarf, is a stellar remnant composed mostly of electron-degenerate matter. A white dwarf is very dense: its mass is comparable to that of the Sun, while its volume is comparable to that of Earth. A white dwarf's faint luminosity comes from the emission of stored thermal energy; no fusion takes place in a white dwarf wherein mass is converted to energy. The nearest known white dwarf is Sirius B, at 8.6 light years, the smaller component of the Sirius binary star. There are currently thought to be eight white dwarfs among the hundred star systems nearest the Sun. The unusual faintness of white dwarfs was first recognized in 1910. The name *white dwarf* was coined by Willem Luyten in 1922. The universe has not existed long enough to experience a white dwarf releasing all of its energy as it will take close to a trillion years.

White dwarfs are thought to be the final evolutionary state of stars whose mass is not high enough to become a neutron star, including the Earth's Sun and over 97% of the other stars in the Milky Way.After the hydrogen-fusing period of a main-sequence star of low or medium mass ends, such a star will expand to a red giant during which it fuses helium to carbon and oxygen in its core by the triple-alpha process. If a red giant has insufficient mass to generate the core temperatures required to fuse carbon, around 1 billion K, an inert mass of carbon and oxygen will build up at its center. After a star sheds its outer layers and forms a planetary nebula, it will leave behind this core, which is the remnant white dwarf. Usually, therefore, white dwarfs are composed of carbon and oxygen. If the mass of the progenitor is between 8 and 10.5 solar masses ($M_\odot$), the core temperature is sufficient to fuse carbon but not neon, in which case an oxygen–neon–magnesium white dwarf may form. Stars of very low mass will not be able to fuse helium, hence, a helium white dwarf may form by mass loss in binary systems.

The material in a white dwarf no longer undergoes fusion reactions, so the star has no source of energy. As a result, it cannot support itself by the heat generated by fusion against gravitational collapse, but is supported only by electron degeneracy pressure, causing it to be extremely dense. The physics of degeneracy yields a maximum mass for a non-rotating white dwarf, the Chandrasekhar limit—approximately 1.46 $M_\odot$—beyond which it cannot be supported by electron degeneracy pressure. A carbon-oxygen white dwarf that approaches this mass limit, typically by mass transfer from a companion star, may explode as a type Ia supernova via a process known as carbon detonation. (SN 1006 is thought to be a famous example.)

A white dwarf is very hot when it forms, but because it has no source of energy, it will gradually radiate its energy and cool. This means that its radiation, which initially has a high color temperature, will lessen and redden with time. Over a very long time, a white dwarf will cool and its material will begin to crystallize (starting with the core). The star's low temperature means it will no longer emit significant heat or light, and it will become a cold *black dwarf*. The length of time it takes for a white dwarf to reach this state is calculated to be longer than the current age of the universe (approximately 13.8 billion years), and it is thought that no black dwarfs yet exist. The oldest white dwarfs still radiate at temperatures of a few thousand kelvins.

Discovery

The first white dwarf discovered was in the triple star system of 40 Eridani, which contains the relatively bright main sequence star 40 Eridani A, orbited at a distance by the closer binary system of the white dwarf 40 Eridani B and the main sequence red dwarf 40 Eridani C. The pair 40 Eridani B/C was discovered by William Herschel on 31 January 1783; it was again observed by Friedrich Georg Wilhelm Struve in 1825 and by Otto Wilhelm von Struve in 1851. In 1910, Henry Norris Russell, Edward Charles Pickering and Williamina Fleming discovered that, despite being a dim star, 40 Eridani B was of spectral type A, or white. In 1939, Russell looked back on the discovery:

I was visiting my friend and generous benefactor, Prof. Edward C. Pickering. With characteristic kindness, he had volunteered to have the spectra observed for all the stars—including comparison stars—which had been observed in the observations for stellar parallax which Hinks and I made at Cambridge, and I discussed. This piece of apparently routine work proved very fruitful—it led to the discovery that all the stars of very faint absolute magnitude were of spectral class M. In conversation on this subject (as I recall it), I asked Pickering about certain other faint stars, not on my list, mentioning in particular 40 Eridani B. Characteristically, he sent a note to the Observatory office and before long the answer came (I think from Mrs Fleming) that the spectrum of this star was A. I knew enough about it, even in these paleozoic days, to realize at once that there was an extreme inconsistency between what we would then have called "possible" values of the surface brightness and density. I must have shown that I was not only puzzled but crestfallen, at this exception to what looked like a very pretty rule of stellar characteristics; but Pickering smiled upon me, and said: "It is just these exceptions that lead to an advance in our knowledge", and so the white dwarfs entered the realm of study!

The spectral type of 40 Eridani B was officially described in 1914 by Walter Adams.

The white dwarf companion of Sirius, Sirius B, was next to be discovered. During the nineteenth century, positional measurements of some stars became precise enough to measure small changes in their location. Friedrich Bessel used position measurements to determine that the stars Sirius (α Canis Majoris) and Procyon (α Canis Minoris) were changing their positions periodically. In 1844 he predicted that both stars had unseen companions:

If we were to regard *Sirius* and *Procyon* as double stars, the change of their motions would not surprise us; we should acknowledge them as necessary, and have only to investigate their amount by observation. But light is no real property of mass. The existence of numberless visible stars can prove nothing against the existence of numberless invisible ones.

Bessel roughly estimated the period of the companion of Sirius to be about half a century; C. A. F. Peters computed an orbit for it in 1851. It was not until 31 January 1862 that Alvan Graham Clark observed a previously unseen star close to Sirius, later identified as the predicted companion. Walter Adams announced in 1915 that he had found the spectrum of Sirius B to be similar to that of Sirius.

In 1917, Adriaan van Maanen discovered Van Maanen's Star, an isolated white dwarf. These three white dwarfs, the first discovered, are the so-called *classical white dwarfs*. Eventually, many faint white stars were found which had high proper motion, indicating that they could be suspected to be low-luminosity stars close to the Earth, and hence white dwarfs. Willem Luyten appears to have been the first to use the term *white dwarf* when he examined this class of stars in 1922; the term was later popularized by Arthur Stanley Eddington. Despite these suspicions, the first non-classical white dwarf was not definitely identified until the 1930s. 18 white dwarfs had been discovered by 1939. Luyten and others continued to search for white dwarfs in the 1940s. By 1950, over a hundred were known, and by 1999, over 2,000 were known. Since then the Sloan Digital Sky Survey has found over 9,000 white dwarfs, mostly new.

Composition and Structure

Although white dwarfs are known with estimated masses as low as $0.17\ M_\odot$ and as high as $1.33\ M_\odot$, the mass distribution is strongly peaked at $0.6\ M_\odot$, and the majority lie between 0.5 and 0.7

$M_\odot$. The estimated radii of observed white dwarfs are typically 0.8–2 % the radius of the Sun; this is comparable to the Earth's radius of approximately 0.9% solar radius. A white dwarf, then, packs mass comparable to the Sun's into a volume that is typically a million times smaller than the Sun's; the average density of matter in a white dwarf must therefore be, very roughly, 1,000,000 times greater than the average density of the Sun, or approximately 10^6 g/cm³, or 1 tonne per cubic centimetre. A typical white dwarf has a density of between 10^4 and 10^7 g/cm^3. White dwarfs are composed of one of the densest forms of matter known, surpassed only by other compact stars such as neutron stars, black holes and, hypothetically, quark stars.

White dwarfs were found to be extremely dense soon after their discovery. If a star is in a binary system, as is the case for Sirius B or 40 Eridani B, it is possible to estimate its mass from observations of the binary orbit. This was done for Sirius B by 1910, yielding a mass estimate of 0.94 $M_\odot$. (A more modern estimate is 1.00 $M_\odot$.) Since hotter bodies radiate more energy than colder ones, a star's surface brightness can be estimated from its effective surface temperature, and that from its spectrum. If the star's distance is known, its absolute (overall) luminosity can also be estimated. From the absolute luminosity and distance, the star's surface area and its radius can be calculated. Reasoning of this sort led to the realization, puzzling to astronomers at the time, that Sirius B and 40 Eridani B must be very dense. When Ernst Öpik estimated the density of a number of visual binary stars in 1916, he found that 40 Eridani B had a density of over 25,000 times the Sun's, which was so high that he called it "impossible". As Arthur Stanley Eddington put it later in 1927:

We learn about the stars by receiving and interpreting the messages which their light brings to us. The message of the Companion of Sirius when it was decoded ran: "I am composed of material 3,000 times denser than anything you have ever come across; a ton of my material would be a little nugget that you could put in a matchbox." What reply can one make to such a message? The reply which most of us made in 1914 was—"Shut up. Don't talk nonsense."

As Eddington pointed out in 1924, densities of this order implied that, according to the theory of general relativity, the light from Sirius B should be gravitationally redshifted. This was confirmed when Adams measured this redshift in 1925.

Material	Density in kg/m³	Notes
Water (fresh)	1,000	At STP
Osmium	22,610	Near room temperature
The core of the Sun	~150,000	
White dwarf	1×10^9	
Atomic nuclei	2.3×10^{17}	Does not depend strongly on size of nucleus
Neutron star core	$8.4 \times 10^{16} - 1 \times 10^{18}$	
Black hole	2×10^{30}	Critical density of an Earth-mass black hole

Such densities are possible because white dwarf material is not composed of atoms joined by chemical bonds, but rather consists of a plasma of unbound nuclei and electrons. There is therefore no obstacle to placing nuclei closer than normally allowed by electron orbitals limited by normal matter. Eddington wondered what would happen when this plasma cooled and the energy to keep the atoms ionized was no longer sufficient. This paradox was resolved by R. H. Fowler in 1926 by an application of the newly devised quantum mechanics. Since electrons obey the Pauli exclusion principle, no two electrons can occupy the same state, and they must obey Fermi–Dirac statistics, also introduced in 1926 to determine the statistical distribution of particles which satisfy the Pauli exclusion principle. At zero temperature, therefore, electrons can not all occupy the lowest-energy, or *ground*, state; some of them would have to occupy higher-energy states, forming a band of lowest-available energy states, the *Fermi sea*. This state of the electrons, called *degenerate*, meant that a white dwarf could cool to zero temperature and still possess high energy.

Compression of a white dwarf will increase the number of electrons in a given volume. Applying the Pauli exclusion principle, this will increase the kinetic energy of the electrons, thereby increasing the pressure. This *electron degeneracy pressure* supports a white dwarf against gravitational collapse. The pressure depends only on density and not on temperature. Degenerate matter is relatively compressible; this means that the density of a high-mass white dwarf is much greater than that of a low-mass white dwarf and that the radius of a white dwarf decreases as its mass increases.

The existence of a limiting mass that no white dwarf can exceed (beyond which it becomes a neutron star) is another consequence of being supported by electron degeneracy pressure. These masses were first published in 1929 by Wilhelm Anderson and in 1930 by Edmund C. Stoner. The modern value of the limit was first published in 1931 by Subrahmanyan Chandrasekhar in his paper "The Maximum Mass of Ideal White Dwarfs". For a non-rotating white dwarf, it is equal to approximately $5.7 M_\odot / \mu_e^2$, where μ_e is the average molecular weight per electron of the star. As the carbon-12 and oxygen-16 which predominantly compose a carbon-oxygen white dwarf both have atomic number equal to half their atomic weight, one should take μ_e equal to 2 for such a star, leading to the commonly quoted value of 1.4 $M_\odot$. (Near the beginning of the 20th century, there was reason to believe that stars were composed chiefly of heavy elements, so, in his 1931 paper, Chandrasekhar set the average molecular weight per electron, μ_e, equal to 2.5, giving a limit of 0.91 $M_\odot$.) Together with William Alfred Fowler, Chandrasekhar received the Nobel prize for this and other work in 1983. The limiting mass is now called the *Chandrasekhar limit*.

If a white dwarf were to exceed the Chandrasekhar limit, and nuclear reactions did not take place, the pressure exerted by electrons would no longer be able to balance the force of gravity, and it would collapse into a denser object called a neutron star. Carbon-oxygen white dwarfs accreting mass from a neighboring star undergo a runaway nuclear fusion reaction, which leads to a Type Ia supernova explosion in which the white dwarf may be destroyed, before it reaches the limiting mass.

New research indicates that many white dwarfs—at least in certain types of galaxies—may not approach that limit by way of accretion. It has been postulated that at least some of the white dwarfs that become supernovae attain the necessary mass by colliding with one another. It may be that in elliptical galaxies such collisions are the major source of supernovae. This hypothesis is based on the fact that the X-rays produced by those galaxies are 30 to 50 times less than what is expected to be produced by type Ia supernovas of that galaxy as matter accretes on the white dwarf from its encircling companion. It has been concluded that no more than 5 percent of the supernovae in such galaxies could be created by the process of accretion onto white dwarfs. The significance of this finding is that there could be two types of supernovae, which could mean that the Chandrasekhar limit might not always apply in determining when a white dwarf goes supernova, given that two colliding white dwarfs could have a range of masses. This in turn would confuse efforts to use exploding white dwarfs as standard candles in determining distances.

White dwarfs have low luminosity and therefore occupy a strip at the bottom of the Hertzsprung–Russell diagram, a graph of stellar luminosity versus color (or temperature). They should not be confused with low-luminosity objects at the low-mass end of the main sequence, such as the hydrogen-fusing red dwarfs, whose cores are supported in part by thermal pressure, or the even lower-temperature brown dwarfs.

Mass–Radius Relationship and Mass Limit

The relationship between the mass and radii of white dwarfs can be derived using an energy minimization argument . The energy of the white dwarf can be approximated by taking it to be the sum of its gravitational potential energy and kinetic energy. The gravitational potential energy of a unit mass piece of white dwarf, E_g, will be on the order of $-GM/R$, where G is the gravitational constant, M is the mass of the white dwarf, and R is its radius.

$$E_g \approx \frac{-GM}{R}.$$

The kinetic energy of the unit mass, E_k, will primarily come from the motion of electrons, so it will be approximately $Np^2/2m$, where p is the average electron momentum, m is the electron mass, and N is the number of electrons per unit mass. Since the electrons are degenerate, we can estimate p to be on the order of the uncertainty in momentum, Δp, given by the uncertainty principle, which says that $\Delta p \Delta x$ is on the order of the reduced Planck constant, $\hbar$. Δx will be on the order of the average distance between electrons, which will be approximately $n^{-1/3}$, i.e., the reciprocal of the cube root of the number density, n, of electrons per unit volume. Since there are $N \cdot M$ electrons in the white dwarf, where M is the star's mass and its volume is on the order of R^3, n will be on the order of NM/R^3.

Solving for the kinetic energy per unit mass, E_k, we find that

$$E_k \approx \frac{N(\Delta p)^2}{2m} \approx \frac{N\hbar^2 n^{2/3}}{2m} \approx \frac{M^{2/3} N^{5/3} \hbar^2}{2mR^2}.$$

The white dwarf will be at equilibrium when its total energy, $E_g + E_k$, is minimized. At this point, the kinetic and gravitational potential energies should be comparable, so we may derive a rough mass-radius relationship by equating their magnitudes:

$$|E_g| \approx \frac{GM}{R} = E_k \approx \frac{M^{2/3} N^{5/3} \hbar^2}{2mR^2}.$$

Solving this for the radius, R, gives

$$R \approx \frac{N^{5/3} \hbar^2}{2mGM^{1/3}}.$$

Dropping N, which depends only on the composition of the white dwarf, and the universal constants leaves us with a relationship between mass and radius:

$$R \sim M^{-1/3}$$

i.e., the radius of a white dwarf is inversely proportional to the cube root of its mass.

Since this analysis uses the non-relativistic formula $p^2/2m$ for the kinetic energy, it is non-relativistic. If we wish to analyze the situation where the electron velocity in a white dwarf is close to the speed of light, c, we should replace $p^2/2m$ by the extreme relativistic approximation pc for the kinetic energy. With this substitution, we find

$$E_{k\,\text{relativistic}} \approx \frac{M^{1/3} N^{4/3} \hbar c}{R}.$$

If we equate this to the magnitude of E_g, we find that R drops out and the mass, M, is forced to be

$$M_{\text{limit}} \approx N^2 \left(\frac{\hbar c}{G} \right)^{3/2}.$$

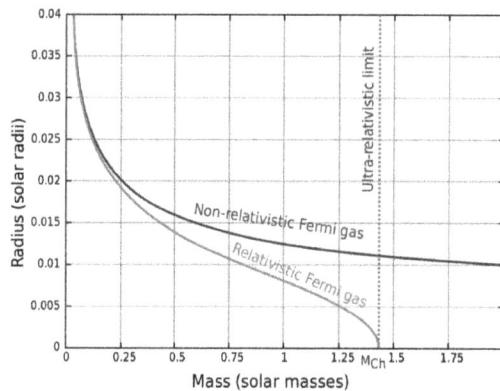

Radius–mass relations for a model white dwarf. *M*limit is denoted as M_{Ch}

To interpret this result, observe that as we add mass to a white dwarf, its radius will decrease, so, by the uncertainty principle, the momentum, and hence the velocity, of its electrons will increase. As this velocity approaches c, the extreme relativistic analysis becomes more exact, meaning that the mass M of the white dwarf must approach a limiting mass of *M*limit. Therefore, no white dwarf can be heavier than the limiting mass *M*limit, or $1.4\ M_\odot$.

For a more accurate computation of the mass-radius relationship and limiting mass of a white dwarf, one must compute the equation of state which describes the relationship between density and pressure in the white dwarf material. If the density and pressure are both set equal to functions of the radius from the center of the star, the system of equations consisting of the hydrostatic equation together with the equation of state can then be solved to find the structure of the white dwarf at equilibrium. In the non-relativistic case, we will still find that the radius is inversely proportional to the cube root of the mass. eq. (80) Relativistic corrections will alter the result so that the radius becomes zero at a finite value of the mass. This is the limiting value of the mass—called the *Chandrasekhar limit*—at which the white dwarf can no longer be supported by electron degeneracy pressure. The graph on the right shows the result of such a computation. It shows how radius varies with mass for non-relativistic (blue curve) and relativistic (green curve) models of a white dwarf. Both models treat the white dwarf as a cold Fermi gas in hydrostatic equilibrium. The average molecular weight per electron, μ_e, has been set equal to 2. Radius is measured in standard solar radii and mass in standard solar masses.

These computations all assume that the white dwarf is non-rotating. If the white dwarf is rotating, the equation of hydrostatic equilibrium must be modified to take into account the centrifugal pseudo-force arising from working in a rotating frame. For a uniformly rotating white dwarf, the limiting mass increases only slightly. If the star is allowed to rotate nonuniformly, and viscosity is neglected, then, as was pointed out by Fred Hoyle in 1947, there is no limit to the mass for which it is possible for a model white dwarf to be in static equilibrium. Not all of these model stars will be dynamically stable.

Radiation and Cooling

The degenerate matter that makes up the bulk of a white dwarf has a very low opacity, because any absorption of a photon requires that an electron must transition to a higher empty state, which may not be possible as the energy of the photon may not be a match for the possible quantum states available to that electron, hence radiative heat transfer within a white dwarf is low; it does, however, have a high thermal conductivity. As a result, the interior of the white dwarf maintains a uniform temperature, approximately 10^7 K. An outer shell of non-degenerate matter cools from approximately 10^7 K to 10^4 K. This matter radiates roughly as a black body. A white dwarf remains visible for a long time, as its tenuous outer atmosphere of normal matter begins to radiate at about 10^7 K, upon formation, while its greater interior mass is at 10^7 K but cannot radiate through its normal matter shell.

The visible radiation emitted by white dwarfs varies over a wide color range, from the blue-white color of an O-type main sequence star to the red of an M-type red dwarf. White dwarf effective surface temperatures extend from over 150,000 K to barely under 4,000 K. In accordance with the Stefan–Boltzmann law, luminosity increases with increasing surface temperature; this surface temperature range corresponds to a luminosity from over 100 times the Sun's to under 1/10,000 that of the Sun's. Hot white dwarfs, with surface temperatures in excess of 30,000 K, have been observed to be sources of soft (i.e., lower-energy) X-rays. This enables the composition and structure of their atmospheres to be studied by soft X-ray and extreme ultraviolet observations.

As was explained by Leon Mestel in 1952, unless the white dwarf accretes matter from a companion star or other source, its radiation comes from its stored heat, which is not replenished. White dwarfs have an extremely small surface area to radiate this heat from, so they cool gradually, remaining hot for a long time. As a white dwarf cools, its surface temperature decreases, the radiation which it emits reddens, and its luminosity decreases. Since the white dwarf has no energy sink other than radiation, it follows that its cooling slows with time. The rate of cooling has been estimated for a carbon white dwarf of 0.59 $M_\odot$ with a hydrogen atmosphere. After initially taking approximately 1.5 billion years to cool to a surface temperature of 7,140 K, cooling approximately 500 more kelvins to 6,590 K takes around 0.3 billion years, but the next two steps of around 500 kelvins (to 6,030 K and 5,550 K) take first 0.4 and then 1.1 billion years.

A comparison between the white dwarf IK Pegasi B (center), its A-class companion IK Pegasi A (left) and the Sun (right). This white dwarf has a surface temperature of 35,500 K.

Most observed white dwarfs have relatively high surface temperatures, between 8,000 K and 40,000 K. A white dwarf, though, spends more of its lifetime at cooler temperatures than at hotter temperatures, so we should expect that there are more cool white dwarfs than hot white dwarfs. Once we adjust for the selection effect that hotter, more luminous white dwarfs are easier to observe, we do find that decreasing the temperature range examined results in finding more white dwarfs. This trend stops when we reach extremely cool white dwarfs; few white dwarfs are observed with surface temperatures below 4,000 K, and one of the coolest so far observed, WD 0346+246, has a surface temperature of approximately 3,900 K. The reason for this is that the Universe's age is finite, there has not been enough time for white dwarfs to cool below this temperature. The white dwarf luminosity function can therefore be used to find the time when stars started to form in a region; an estimate for the age of our Galactic disk found in this way is 8 billion years. A white dwarf will eventually, in many trillion years, cool and become a non-radiating *black dwarf* in approximate thermal equilibrium with its surroundings and with the cosmic background radiation. No black dwarfs are thought to exist yet.

Although white dwarf material is initially plasma—a fluid composed of nuclei and electrons—it was theoretically predicted in the 1960s that at a late stage of cooling, it should crystallize, starting at its center. The crystal structure is thought to be a body-centered cubic lattice. In 1995 it was suggested that asteroseismological observations of pulsating white dwarfs yielded a potential test of the crystallization theory, and in 2004, observations were made that suggested approximately 90% of the mass of BPM 37093 had crystallized. Other work gives a crystallized mass fraction of between 32% and 82%. As a white dwarf core undergoes crystallization into a solid phase, latent heat is released which provides a source of thermal energy that delays its cooling.

Low-mass helium white dwarfs (with a mass $< 0.20 \, M_\odot$, often referred to as "extremely low-mass white dwarfs, ELM WDs") are formed in binary systems. As a result of their hydrogen-rich envelopes, residual hydrogen burning via the CNO cycle may keep these white dwarfs hot on a long timescale. In addition, they remain in a bloated proto-white dwarf stage for up to 2 Gyr before they reach the cooling track.

Atmosphere and Spectra

Although most white dwarfs are thought to be composed of carbon and oxygen, spectroscopy typically shows that their emitted light comes from an atmosphere which is observed to be either hydrogen or helium dominated. The dominant element is usually at least 1,000 times more abundant than all other elements. As explained by Schatzman in the 1940s, the high surface gravity is thought to cause this purity by gravitationally separating the atmosphere so that heavy elements are below and the lighter above. This atmosphere, the only part of the white dwarf visible to us, is thought to be the top of an envelope which is a residue of the star's envelope in the AGB phase and may also contain material accreted from the interstellar medium. The envelope is believed to consist of a helium-rich layer with mass no more than 1/100 of the star's total mass, which, if the atmosphere is hydrogen-dominated, is overlain by a hydrogen-rich layer with mass approximately 1/10,000 of the stars total mass.·

Although thin, these outer layers determine the thermal evolution of the white dwarf. The degenerate electrons in the bulk of a white dwarf conduct heat well. Most of a white dwarf's mass is therefore at almost the same temperature (isothermal), and it is also hot: a white dwarf with surface temperature between 8,000 K and 16,000 K will have a core temperature between approximately 5,000,000 K and 20,000,000 K. The white dwarf is kept from cooling very quickly only by its outer layers' opacity to radiation.

White dwarf spectral types	
Primary and secondary features	
A	H lines present; no He I or metal lines
B	He I lines; no H or metal lines
C	Continuous spectrum; no lines
O	He II lines, accompanied by He I or H lines
Z	Metal lines; no H or He I lines
Q	Carbon lines present
X	Unclear or unclassifiable spectrum
Secondary features only	
P	Magnetic white dwarf with detectable polarization
H	Magnetic white dwarf without detectable polarization
E	Emission lines present
V	Variable

The first attempt to classify white dwarf spectra appears to have been by G. P. Kuiper in 1941, and various classification schemes have been proposed and used since then. The system currently in use was introduced by Edward M. Sion, Jesse L. Greenstein and their coauthors in 1983 and has been subsequently revised several times. It classifies a spectrum by a symbol which consists of an initial D, a letter describing the primary feature of the spectrum followed by an optional sequence of letters describing secondary features of the spectrum (as shown in the table to the right), and a temperature index number, computed by dividing 50,400 K by the effective temperature. For example:

A white dwarf with only He I lines in its spectrum and an effective temperature of 15,000 K could be given the classification of DB3, or, if warranted by the precision of the temperature measurement, DB3.5.

A white dwarf with a polarized magnetic field, an effective temperature of 17,000 K, and a spectrum dominated by He I lines which also had hydrogen features could be given the classification of DBAP3.

The symbols ? and : may also be used if the correct classification is uncertain.

White dwarfs whose primary spectral classification is DA have hydrogen-dominated atmospheres. They make up the majority (approximately 80%) of all observed white dwarfs. The next class in number is of DBs (approximately 16%). The hot (above 15,000 K) DQ class (roughly 0.1%) have carbon-dominated atmospheres. Those classified as DB, DC, DO, DZ, and cool DQ have helium-dominated atmospheres. Assuming that carbon and metals are not present, which spectral classification is seen depends on the effective temperature. Between approximately 100,000 K to 45,000 K, the spectrum will be classified DO, dominated by singly ionized helium. From 30,000 K to 12,000 K, the spectrum will be DB, showing neutral helium lines, and below about 12,000 K, the spectrum will be featureless and classified DC.

Molecular hydrogen (H_2) has been detected in spectra of the atmospheres of some white dwarfs.

Metal-Rich White Dwarfs

Around 25-33% of white dwarfs have metal lines in their spectra, which is unusual because any heavy elements in a white dwarf should sink into the star's interior in just a small fraction of the star's lifetime. The prevailing explanation for metal-rich white dwarfs is that they have recently accreted rocky planetesimals. The bulk composition of the accreted object can be measured from the strengths of the metal lines. For example, a 2015 study of the white dwarf Ton 345 concluded that its metal abundances were consistent with those of a differentiated, rocky planet whose mantle had been eroded by the host star's wind during its asymptotic giant branch phase.

Magnetic Field

Magnetic fields in white dwarfs with a strength at the surface of ~1 million gauss (100 teslas) were predicted by P. M. S. Blackett in 1947 as a consequence of a physical law he had proposed which stated that an uncharged, rotating body should generate a magnetic field proportional to its angular momentum. This putative law, sometimes called the *Blackett effect*, was never generally accepted, and by the 1950s even Blackett felt it had been refuted. In the 1960s, it was proposed that white dwarfs might have magnetic fields due to conservation of total surface magnetic flux that existed in its progenitor star phase. A surface magnetic field of ~100 gauss (0.01 T) in the progenitor star would thus become a surface magnetic field of ~100·100^2 = 1 million gauss (100 T) once the star's radius had shrunk by a factor of 100. The first magnetic white dwarf to be observed was GJ 742, which was detected to have a magnetic field in 1970 by its emission of circularly polarized light. It is thought to have a surface field of approximately 300 million gauss (30 kT). Since then magnetic fields have been discovered in well over 100 white dwarfs, ranging from 2×10^3 to 10^9 gauss (0.2 T to 100 kT). Only a small number of white dwarfs have been examined for fields, and it has been estimated that at least 10% of white dwarfs have fields in excess of 1 million gauss (100 T).

Chemical Bonds

The magnetic fields in a white dwarf may allow for the existence of a new type of chemical bond, perpendicular paramagnetic bonding, in addition to ionic and covalent bonds, resulting in what has been initially described as "magnetized matter" in research published in 2012.

Variability

DAV (GCVS: *ZZA*)	DA spectral type, having only hydrogen absorption lines in its spectrum
DBV (GCVS: *ZZB*)	DB spectral type, having only helium absorption lines in its spectrum
GW Vir (GCVS: *ZZO*)	Atmosphere mostly C, He and O; may be divided into DOV and PNNV stars
Types of pulsating white dwarf	

Early calculations suggested that there might be white dwarfs whose luminosity varied with a period of around 10 seconds, but searches in the 1960s failed to observe this. The first variable white dwarf found was HL Tau 76; in 1965 and 1966, and was observed to vary with a period of approximately 12.5 minutes. The reason for this period being longer than predicted is that the variability of HL Tau 76, like that of the other pulsating variable white dwarfs known, arises from non-radial gravity wave pulsations. Known types of pulsating white dwarf include the *DAV*, or *ZZ Ceti*, stars, including HL Tau 76, with hydrogen-dominated atmospheres and the spectral type DA;*DBV*, or *V777 Her*, stars, with helium-dominated atmospheres and the spectral type DB; and *GW Vir stars* (sometimes subdivided into *DOV* and *PNNV* stars), with atmospheres dominated by helium, carbon, and oxygen. GW Vir stars are not, strictly speaking, white dwarfs, but are stars which are in a position on the Hertzsprung-Russell diagram between the asymptotic giant branch and the white dwarf region. They may be called *pre-white dwarfs*. These variables all exhibit small (1%–30%) variations in light output, arising from a superposition of vibrational modes with periods of hundreds to thousands of seconds. Observation of these variations gives asteroseismological evidence about the interiors of white dwarfs.

Formation

White dwarfs are thought to represent the end point of stellar evolution for main-sequence stars with masses from about 0.07 to 10 $M_\odot$. The composition of the white dwarf produced will depend on the initial mass of the star.

Stars with very Low Mass

If the mass of a main-sequence star is lower than approximately half a solar mass, it will never become hot enough to fuse helium in its core. It is thought that, over a lifespan that considerably exceeds the age of the Universe (~13.8 billion years), such a star will eventually burn all its hydrogen and end its evolution as a helium white dwarf composed chiefly of helium-4 nuclei. Due to the very long time this process takes, it is not thought to be the origin of the observed helium white dwarfs. Rather, they are thought to be the product of mass loss in binary systems or mass loss due to a large planetary companion.

Stars with low to Medium Mass

If the mass of a main-sequence star is between 0.5 and 8 $M_\odot$, its core will become sufficiently hot to fuse helium into carbon and oxygen via the triple-alpha process, but it will never become sufficiently hot to fuse carbon into neon. Near the end of the period in which it undergoes fusion reactions, such a star will have a carbon-oxygen core which does not undergo fusion reactions, surrounded by an inner helium-burning shell and an outer hydrogen-burning shell. On the Hertzsprung-Russell diagram, it will be found on the asymptotic giant branch. It will then expel most of its outer material, creating a planetary nebula, until only the carbon-oxygen core is left. This process is responsible for the carbon-oxygen white dwarfs which form the vast majority of observed white dwarfs.

Stars with Medium to High Mass

If a star is massive enough, its core will eventually become sufficiently hot to fuse carbon to neon, and then to fuse neon to iron. Such a star will not become a white dwarf, because the mass of its central, non-fusing core, initially supported by electron degeneracy pressure, will eventually exceed the largest possible mass supportable by degeneracy pressure. At this point the core of the star will collapse and it will explode in a core-collapse supernova which will leave behind a remnant neutron star, black hole, or possibly a more exotic form of compact star. Some main-sequence stars, of perhaps 8 to 10 $M_{\odot}$, although sufficiently massive to fuse carbon to neon and magnesium, may be insufficiently massive to fuse neon. Such a star may leave a remnant white dwarf composed chiefly of oxygen, neon, and magnesium, provided that its core does not collapse, and provided that fusion does not proceed so violently as to blow apart the star in a supernova Although a few white dwarfs have been identified which may be of this type, most evidence for the existence of such comes from the novae called *ONeMg* or *neon* novae. The spectra of these novae exhibit abundances of neon, magnesium, and other intermediate-mass elements which appear to be only explicable by the accretion of material onto an oxygen-neon-magnesium white dwarf.

Type Ia Supernovae

Type Ia supernovae, that involve one or two previous white dwarfs, have been proposed to be a channel for transformation of this type of stellar remnant. In this scenario, the carbon detonation produced in a Type Ia supernova is too weak to destroy the white dwarf, expelling just a small part of its mass as ejecta, but produces an asymmetric explosion that kicks the star to high speeds of a Hypervelocity star. The matter processed in the failed detonation is re-accreted by the white dwarf with the heaviest elements such as iron falling to its core where it accumulates.

These *iron-core* white dwarfs would be smaller than their carbon-oxygen kind of similar mass and would cool and crystallize faster than those.

Fate

Artist's impression of debris around a white dwarf

A white dwarf is stable once formed and will continue to cool almost indefinitely, eventually to become a black dwarf. Assuming that the Universe continues to expand, it is thought that in 10^{19} to 10^{20} years, the galaxies will evaporate as their stars escape into intergalactic space. White dwarfs should generally survive galactic dispersion, although an occasional collision between white dwarfs may produce a new fusing star or a super-Chandrasekhar mass white dwarf which will explode in a Type Ia supernova., §IIIC, IV. The subsequent lifetime of white dwarfs is thought to be on the order of the lifetime of the proton, known to be at least 10^{34}-10^{35} years. Some grand unified theories predict a proton lifetime between 10^{30} and 10^{36} years. If these theories are not valid, the proton may decay by complicated nuclear reactions or through quantum gravitational processes involving a virtual black hole; in these cases, the lifetime is estimated to be no more than 10^{200} years. If protons do decay, the mass of a white dwarf will decrease very slowly with time as its nuclei decay, until it loses enough mass to become a nondegenerate lump of matter, and finally disappears completely.

A white dwarf can also be cannibalized or evaporated by a companion star, causing the white dwarf to lose so much mass that it becomes a planetary mass object. The resultant object, orbiting the former companion, now host star, could be a helium planet or diamond planet.

Debris Disks and Planets

The merger process of two co-orbiting white dwarfs produces gravitational waves

A white dwarf's stellar and planetary system is inherited from its progenitor star and may interact with the white dwarf in various ways. Infrared spectroscopic observations made by NASA's Spitzer Space Telescope of the central star of the Helix Nebula suggest the presence of a dust cloud, which may be caused by cometary collisions. It is possible that infalling material from this may cause X-ray emission from the central star. Similarly, observations made in 2004 indicated the presence of a dust cloud around the young white dwarf G29-38 (estimated to have formed from its AGB progenitor about 500 million years ago), which may have been created by tidal disruption of a comet passing close to the white dwarf. Some estimations based on the metal content of the atmospheres of the white dwarfs consider that at least a 15% of them may be orbited by planets and/or asteroids, or at least their debris. Another suggested idea is that white dwarfs could be orbited by the stripped cores of rocky planets, that would have survived the red giant phase of their star but losing their outer layers and, given those planetary remnants would likely be made of metals, to attempt to detect them looking for the signatures of their interaction with the white dwarf's magnetic field.

There is a planet in the white dwarf–pulsar binary system PSR B1620-26.

There are two circumbinary planets around the white dwarf–red dwarf binary NN Serpentis.

The metal-rich white dwarf WD 1145+017 is the first white dwarf observed with a disintegrating minor planet which transits the star. The disintegration of the planetesimal generates a debris cloud which passes in front of the star every 4.5 hours, causing a 5-minute-long fade in the star's optical brightness. The depth of the transit is highly variable.

Habitability

It has been proposed that white dwarfs with surface temperatures of less than 10,000 kelvins could harbor a habitable zone at a distance between ~0.005 to 0.02 AU that would last upwards of 3 billion years. The goal is to search for transits of hypothetical Earth-like planets that could have migrated inward and/or formed there. As a white dwarf has a size similar to that of a planet, these kinds of transits would produce strong eclipses. Newer research casts some doubts on this idea, given that the close orbits of those hypothetical planets around their parent stars would subject them to strong tidal forces that could render them unhabitable by triggering a greenhouse effect. Another suggested constraint to this idea is the origin of those planets. Leaving aside in-situ formation on an accretion disk surrounding the white dwarf, there are two ways a planet could end in a close orbit around stars of this kind: by surviving being engulfed by the star during its red giant phase, and then spiraling towards its core, or inward migration after the white dwarf has formed. The former case is implausible for low-mass bodies, as they are unlikely to survive being absorbed by their stars. In the latter case, the planets would have to expel so much orbital energy as heat, through tidal interactions with the white dwarf, that they would likely end as uninhabitable embers.

Binary Stars and Novae

If a white dwarf is in a binary star system and is accreting matter from its companion, a variety of phenomena may occur, including novae and Type Ia supernovae. It may also be a super-soft x-ray source if it is able to take material from its companion fast enough to sustain fusion on its surface. A close binary system of two white dwarfs can radiate energy in the form of gravitational waves, causing their mutual orbit to steadily shrink until the stars merge.

Type Ia Supernovae

The mass of an isolated, nonrotating white dwarf cannot exceed the Chandrasekhar limit of ~1.4 $M_\odot$. (This limit may increase if the white dwarf is rotating rapidly and nonuniformly.) White dwarfs in binary systems can accrete material from a companion star, increasing both their mass and their density. As their mass approaches the Chandrasekhar limit, this could theoretically lead to either the explosive ignition of fusion in the white dwarf or its collapse into a neutron star.

Accretion provides the currently favored mechanism called the *single-degenerate model* for Type Ia supernovae. In this model, a carbon–oxygen white dwarf accretes mass and compresses its core by pulling mass from a companion star. It is believed that compressional heating of the core leads to ignition of carbon fusion as the mass approaches the Chandrasekhar limit. Because the white dwarf is supported against gravity by quantum degeneracy pressure instead of by thermal pressure, adding heat to the star's interior increases its temperature but not its pressure, so the white dwarf does not expand and cool in response. Rather, the increased temperature accelerates the rate of the fusion reaction, in a runaway process that feeds on itself. The thermonuclear flame consumes much of the white dwarf in a few seconds, causing a Type Ia supernova explosion that obliterates the star. In another possible mechanism for Type Ia supernovae, the *double-degenerate model*, two carbon-oxygen white dwarfs in a binary system merge, creating an object with mass greater than the Chandrasekhar limit in which carbon fusion is then ignited.

Observations have failed to note signs of accretion leading up to Type Ia supernovae, and this is now thought to be because the star is first loaded up to above the Chandrasekhar limit while also being spun up to a very high rate by the same process. Once the accretion stops the star gradually slows until the spin is no longer enough to prevent the explosion.

Cataclysmic Variables

Before accretion of material pushes a white dwarf close to the Chandrasekhar limit, accreted hydrogen-rich material on the surface may ignite in a less destructive type of thermonuclear explosion powered by hydrogen fusion. These surface explosions can be repeated as long as the white dwarf's core remains intact. This weaker kind of repetitive cataclysmic phenomenon is called a (classical) nova. Astronomers have also observed dwarf novae, which have smaller, more frequent luminosity peaks than the classical novae. These are thought to be caused by the release of gravitational potential energy when part of the accretion disc collapses onto the star, rather than through a release of energy due to fusion. In general, binary systems with a white dwarf accreting matter from a stellar companion are called cataclysmic variables. As well as novae and dwarf novae, several other classes of these variables are known, including polars and intermediate polars, both of which feature highly magnetic white dwarfs. Both fusion- and accretion-powered cataclysmic variables have been observed to be X-ray sources.

Nearest

White Dwarfs within 25 Light Years							
Identifier	WD Number	Distance (ly)	Type	Absolute magnitude	Mass $(M_\odot)$	Luminosity $(L_\odot)$	Age (Gyr)
Sirius B	0642–166	8.61	DA	11.18	0.98	0.0295	0.10
Procyon B 0736+053		11.4	DQZ	13.20	0.63	0.00049	1.37

Van Maanen 2	0046+051	14.4	DZ	14.09	0.68	0.00017	3.30
LP 145-141	1142−645	15.1	DQ	12.77	0.61	0.00054	1.29
40 Eridani B	0413-077	16.6	DA	11.27	0.59	0.0141	0.12
Stein 2051 B	0426+588	18.0	DC	13.43	0.69	0.00030	2.02
L 97-12	0552−041	21.1	DZ	15.29	0.82	0.000062	7.89

Red Giant

A red giant is a luminous giant star of low or intermediate mass (roughly 0.3–8 solar masses ($M_\odot$)) in a late phase of stellar evolution. The outer atmosphere is inflated and tenuous, making the radius large and the surface temperature as low as 5,000 K and lower. The appearance of the red giant is from yellow-orange to red, including the spectral types K and M, but also class S stars and most carbon stars.

The most common red giants are stars on the red-giant branch (RGB) that are still fusing hydrogen into helium in a shell surrounding an inert helium core. Other red giants are the red-clump stars in the cool half of the horizontal branch, fusing helium into carbon in their cores via the triple-alpha process; and the asymptotic-giant-branch (AGB) stars with a helium burning shell outside a degenerate carbon–oxygen core, and a hydrogen burning shell just beyond that.

Characteristics

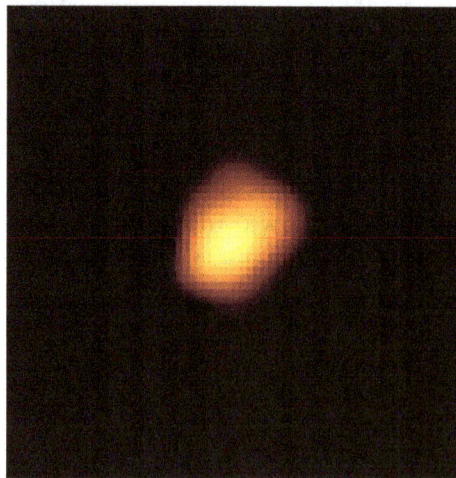

Mira, a variable asymptotic giant branch red giant

Red giants are stars that have exhausted the supply of hydrogen in their cores and have begun thermonuclear fusion of hydrogen in a shell surrounding the core. They have radii tens to hundreds of times larger than that of the Sun. However, their outer envelope is lower in temperature, giving them a reddish-orange hue. Despite the lower energy density of their envelope, red giants are many times more luminous than the Sun because of their great size. Red-giant-branch stars have luminosities up to nearly three thousand times that of the Sun ($L_\odot$), spectral types of K or M, have surface temperatures of 3,000–4,000 K, and radii up to about 200 times the Sun ($R_\odot$). Stars on the horizontal branch are hotter, with only a small range of luminosities around 75 $L_\odot$. Asymptotic-giant-branch stars range from similar luminosities as the brighter stars of the red giant branch, up to several times more luminous at the end of the thermal pulsing phase.

Among the asymptotic-giant-branch stars belong the carbon stars of type C-N and late C-R, produced when carbon and other elements are convected to the surface in what is called a dredge-up. The first dredge-up occurs during hydrogen shell burning on the red-giant branch, but does not produce a large carbon abundance at the surface. The second, and sometimes third, dredge up occurs during helium shell burning on the asymptotic-giant branch and convects carbon to the surface in sufficiently massive stars.

The stellar limb of a red giant is not sharply-defined, contrary to their depiction in many illustrations. Rather, due to the very low mass density of the envelope, such stars lack a well-defined photosphere, and the body of the star gradually transitions into a 'corona'. The coolest red giants have complex spectra, with molecular lines, emission features, and sometimes masers, particularly from thermally pulsing AGB stars.

Another noteworthy feature of red giants is that, unlike Sun-like stars whose photospheres have a large number of small convection cells (solar granules), red-giant photospheres, as well as those of red supergiants, have just a few large cells, whose feature cause the variations of brightness so common on both types of stars.

Evolution

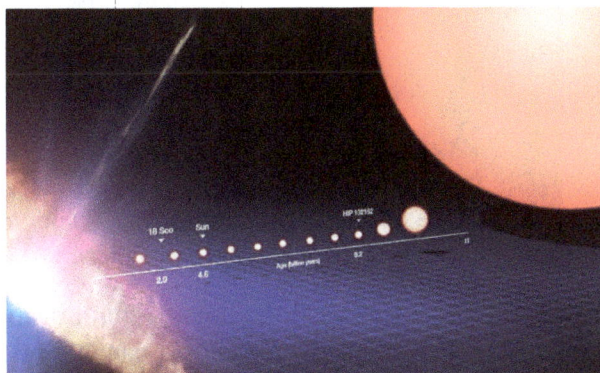

This image tracks the life of a Sun-like star, from its birth on the *left* side of the frame to its evolution into a red giant on the *right* after billions of years.

Red giants are evolved from main-sequence stars with masses in the range from about 0.3 $M_\odot$ to around 8 $M_\odot$. When a star initially forms from a collapsing molecular cloud in the interstellar medium, it contains primarily hydrogen and helium, with trace amounts of "metals" (in stellar structure, this simply refers to *any* element that is not hydrogen or helium i.e. atomic number greater than 2). These elements are all uniformly mixed throughout the star. The star reaches the main sequence when the core reaches a temperature high enough to begin fusing hydrogen (a few million kelvin) and establishes hydrostatic equilibrium. Over its main sequence life, the star slowly converts the hydrogen in the core into helium; its main-sequence life ends when nearly all the hydrogen in the core has been fused. For the Sun, the main-sequence lifetime is approximately 10 billion years. More-massive stars burn disproportionately faster and so have a shorter lifetime than less massive stars.

When the star exhausts the hydrogen fuel in its core, nuclear reactions can no longer continue and so the core begins to contract due to its own gravity. This brings additional hydrogen into a zone where the temperature and pressure are adequate to cause fusion to resume in a shell around the core. The higher temperatures lead to increasing reaction rates, enough to increase the star's luminosity by a factor of 1,000–10,000. The outer layers of the star then expand greatly, thus beginning the red-giant phase of the star's life. As the star expands, the energy produced in the burning shell of the star is spread over a much larger surface area, resulting in a lower surface temperature and a shift in the star's visible light output towards the red – hence it becomes a *red giant*. In actuality, though, the color usually is orange. At this time, the star is said to be ascending the red-giant branch of the Hertzsprung–Russell (H–R) diagram. The outer layers carry the energy evolved from fusion to the surface by way of convection. This causes material exposed to nuclear "burning" in the star's interior (but not its core) to be brought to the star's surface for the first time in its history, an event called the first dredge-up.

Mira A is an old star, already shedding its outer layers into space.

The evolutionary path the star takes as it moves along the red-giant branch, that ends finally with the complete collapse of the core, depends on the mass of the star. For the Sun and stars of less than about 2 $M_\odot$ the core will become dense enough that electron degeneracy pressure will prevent it from collapsing further. Once the core is degenerate, it will continue to heat until it reaches a temperature of roughly 10^8 K, hot enough to begin fusing helium to carbon via the triple-alpha process. Once the degenerate core reaches this temperature, the entire core will begin helium fusion nearly simultaneously in a so-called helium flash. In more-massive stars, the collapsing core will reach 10^8 K before it is dense enough to be degenerate, so helium fusion will begin much more smoothly, and produce no helium flash. Once the star is fusing helium in its core, it contracts and is no longer considered a red giant. The core helium fusing phase of a star's life is called the horizontal branch in metal-poor stars, so named because these stars lie on a nearly horizontal line in the H–R diagram of many star clusters. Metal-rich helium-fusing stars instead lie on the so-called red clump in the H–R diagram.

In stars massive enough to ignite helium fusion, an analogous process occurs when the central helium is exhausted and the star collapses once again, causing helium in an outer shell to begin fusing. At the same time hydrogen may begin fusion in a shell just outside the burning helium shell. This puts the star onto the asymptotic giant branch, a second red-giant phase. The helium fusion results in the build up of a carbon–oxygen core. A star below about 8 $M_\odot$ will never start fusion in its degenerate carbon–oxygen core. Instead, at the end of the asymptotic-giant-branch phase the star will eject its outer layers, forming a planetary nebula with the core of the star exposed, ultimately becoming a white dwarf. The ejection of the outer mass and the creation of a planetary nebula finally ends the red-giant phase of the star's evolution. The red-giant phase typically lasts only around a billion years in total for a solar mass star, almost all of which is spent on the red-giant branch. The horizontal-branch and asymptotic-giant-branch phases proceed tens of times faster.

If the star has about 0.2 to 0.5 $M_\odot$, it is massive enough to become a red giant but does not have enough mass to initiate the fusion of helium. These "intermediate" stars cool somewhat and increase their luminosity but never achieve the tip of the red-giant branch and helium core flash. When the ascent of the red-giant branch ends they puff off their outer layers much like a post-asymptotic-giant-branch star and then become a white dwarf.

Stars that do not become Red Giants

Very low mass stars are fully convective and may continue to fuse hydrogen into helium for up to a trillion years until only a small fraction of the entire star is hydrogen. Luminosity and temperature steadily increase during this time, just as for more-massive main-sequence stars, but the length of time involved means that the temperature eventually increases by about 50% and the luminosity by around 10 times. Eventually the level of helium increases to the point where the star ceases to be fully convective and the remaining hydrogen locked in the core is consumed in only a few billion more years. Depending on mass, the temperature and luminosity continue to increase for a time during hydrogen shell burning, the star can become hotter than the Sun and tens of times more luminous than when it formed although still not as luminous as the Sun. After some billions more years, they start to become less luminous and cooler even though hydrogen shell burning continues. These become cool helium white dwarfs.

Very-high-mass stars develop into supergiants that follow an evolutionary track that takes them back and forth horizontally over the HR diagram, at the right end constituting red supergiants. These usually end their life as a type II supernova. The most massive stars can become Wolf–Rayet stars without becoming giants or supergiants at all.

Planets

Red giants with known planets: the M-type HD 208527, HD 220074 and, as of February 2014, a few tens of known K-giants including Pollux, Gamma Cephei and Iota Draconis.

Prospects for Habitability

Although traditionally it has been suggested the evolution of a star into a red giant will render its planetary system, if present, uninhabitable, some research suggests that, during the evolution of a 1 $M_\odot$ star along the red-giant branch, it could harbor a habitable zone for several times 10^9 years at 2 AU out to around 10^8 years at 9 AU out, giving perhaps enough time for life to develop on a suitable world. After the red-giant stage, there would for such a star be a habitable zone between 7 and 22 AU for an additional 10^9 years. Later studies have refined this scenario, showing how for a 1 $M_\odot$ star the habitable zone lasts from 10^8 years for a planet with an orbit similar to that of Mars one to 2.1×10^8 yr for one that orbits at Saturn's distance to the Sun, the maximum time (3.7×10^8 yr) corresponding for planets orbiting at the distance of Jupiter. However, for planets orbiting a 0.5 $M_\odot$ star in equivalent orbits to those of Ju-piter and Saturn they would be in the habitable zone for 5.8×10^9 yr and 2.1×10^9 yr respectively; for stars more massive than the Sun, the times are considerably shorter.

Enlargement of Planets

As of June 2014, 50 giant planets have been discovered around giant stars. However, these giant planets are more massive than the giant planets found around solar-type stars. This could be because giant stars are more massive than the Sun (less massive stars will still be on the main sequence and will not have become giants yet) and more massive stars are expected to have more massive planets. However, the masses of the planets that have been found around giant stars do not correlate with the masses of the stars; therefore, the planets could be growing in mass during the stars' red giant phase. The growth in planet mass could be partly due to accretion from stellar wind, although a much larger effect would be Roche lobe overflow causing mass-transfer from the star to the planet when the giant expands out to the orbital distance of the planet.

Well Known Examples

Many of the well known bright stars are red giants, because they are luminous and moderately common. The asymptotic giant branch variable star Gamma Crucis is the nearest M class giant star at 88 light years. The K0 red giant branch star Arcturus is 36 light years away.

Red-Giant Branch

- Aldebaran (α Tauri)

- Arcturus (α Bootis)

- Gacrux (γ Crucis)

Red-Clump Giants

- Hamal (α Arietis)

- κ Persei

- δ Andromedae

Asymptotic Giant Branch

- Mira (o Ceti)

- χ Cygni

- α Herculis

The Sun as a Red Giant

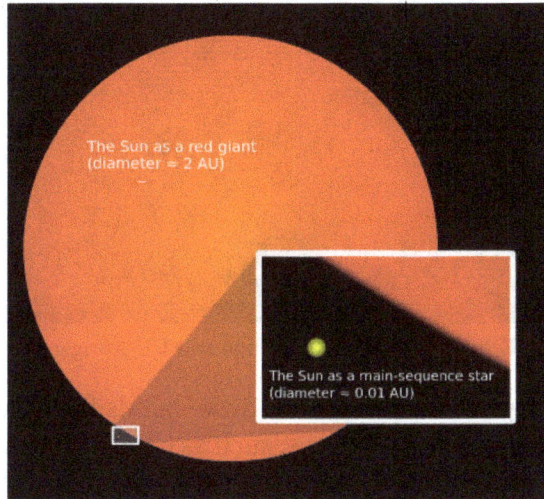

The current size of the Sun (now in the main sequence) compared to its estimated maximum size during its red-giant phase in the future

In about 5 to 6 billion years, the Sun will have depleted the hydrogen fuel in its core and will begin to expand. At its largest, its surface (photosphere) will approximately reach the current orbit of Earth. It will then lose its atmosphere completely; its outer layers forming a planetary nebula and the core a white dwarf. The evolution of the Sun into and through the red-giant phase has been extensively modelled, but it remains unclear whether Earth will be engulfed by the Sun or will continue in orbit. The uncertainty arises in part because as the Sun burns hydrogen, it loses mass causing Earth (and all planets) to orbit farther away. There are also significant uncertainties in calculating the orbits of the planets over the next 5–6.5 billion years, so the fate of Earth is not well understood. At its brightest, the red-giant Sun will be several thousand times more luminous than today but its surface will be at about half the temperature. In its red giant phase, the Sun will be so bright that any water on Earth will boil away into space, leaving it unable to support life.

Carbon Star

A carbon star is a late-type star similar to a red giant (or occasionally to a red dwarf) whose atmosphere contains more carbon than oxygen; the two elements combine in the upper layers of the star, forming carbon monoxide, which consumes all the oxygen in the atmosphere, leaving carbon atoms free to form other carbon compounds, giving the star a "sooty" atmosphere and a strikingly ruby red appearance.

In most stars (such as the Sun), the atmosphere is richer in oxygen than carbon. Ordinary stars not exhibiting the characteristics of carbon stars but cool enough to form carbon monoxide are therefore called oxygen-rich stars.

Carbon stars have quite distinctive spectral characteristics, and they were first recognized by their spectra by Angelo Secchi in the 1860s, a pioneering time in astronomical spectroscopy.

Carbon star Spectra

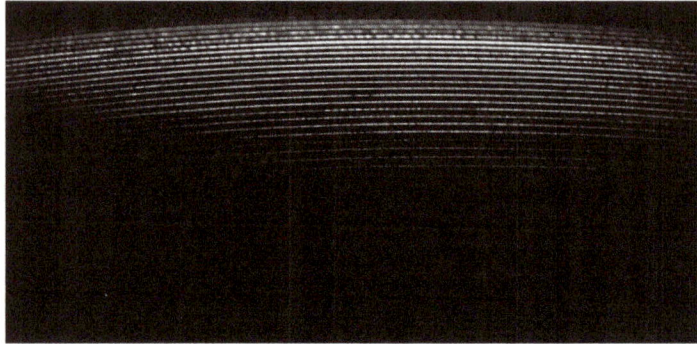

Echelle spectra of the carbon star UU Aurigae

By definition carbon stars have dominant spectral Swan bands from the molecule C_2. Many other carbon compounds may be present at high levels, such as CH, CN (cyanogen), C_3 and SiC_2. Carbon is formed in the core and circulated into its upper layers, dramatically changing the layers' composition. Other elements formed through helium fusion and the s-process are also "dredged up" in this way, including lithium and barium.

When astronomers developed the spectral classification of the carbon stars, they had considerable difficulty when trying to correlate the spectra to the stars' effective temperatures. The trouble was with all the atmospheric carbon hiding the absorption lines normally used as temperature indicators for the stars.

Carbon stars also show a rich spectrum of molecular lines at millimeter wavelengths and submillimeter wavelengths. In the carbon star IRC+10216 more than 50 different circumstellar molecules have been detected. This star is often used to search for new circumstellar molecules.

Secchi

Carbon stars were discovered already in the 1860s when spectral classification pioneer Pater Angelo Secchi erected the Secchi class IV for the carbon stars, which in the late 1890s were reclassified as N class stars.

Harvard

Using this new Harvard classification, the N class was later enhanced by an R class for less deeply red stars sharing the characteristic carbon bands of the spectrum. Later correlation of this R to N scheme with conventional spectra, showed that the R-N sequence approximately run in parallel with c:a G7 to M10 with regards to star temperature.

MK-type	R0	R3	R5	R8	Na	Nb
giant equiv.	G7-G8	K1-K2	~K2-K3	K5-M0	~M2-M3	M3-M4
T_{eff}	4300	3900	~3700	3450	---	---

Morgan–Keenan C System

The later N classes correspond less well to the counterparting M types, because the Harvard classification was only partially based on temperature, but also carbon abundance; so it soon became clear that this kind of carbon star classification was incomplete. Instead a new dual number star class C was erected so to deal with temperature and carbon abundance. Such a spectrum measured for Y Canum Venaticorum, was determined to be $C5_4$, where 5 refers to temperature dependent features, and 4 to the strength of the C_2 Swan bands in the spectrum. ($C5_4$ is very often alternatively written C5,4). This Morgan–Keenan C system classification replaced the older R-N classifications from 1960–1993.

MK-type	C0	C1	C2	C3	C4	C5	C6	C7
giant equiv.	G4-G6	G7-G8	G9-K0	K1-K2	K3-K4	K5-M0	M1-M2	M3-M4
T_{eff}	4500	4300	4100	3900	3650	3450	---	---

The Revised Morgan–Keenan System

The two-dimensional Morgan–Keenan C classification failed to fulfill the creators' expectations:

- it failed to correlate to temperature measurements based on infrared,

- originally being two-dimensional it was soon enhanced by suffixes, CH, CN, j and other features making it impractical for en-masse analyses of foreign galaxies' carbon star populations,

- and it gradually occurred that the old R and N stars actually were two distinct types of carbon stars, having real astrophysical significance.

A new revised Morgan–Keenan classification was published in 1993 by Philip Keenan, defining the classes: C-N, C-R and C-H. Later the classes C-J and C-Hd were added. This constitutes the established classification system used today.

class	spectrum	population	M_v	theory	temperature range (K)	example(s)	# known
classical carbon stars							
C-R:	the old Harvard class R reborn: are still visible at the blue end of the spectrum, strong isotopic bands, no enhanced Ba line	medium disc pop I	0	red giants?	5100-2800	S Cam	~25
C-N:	the old Harvard class N reborn: heavy diffuse blue absorption, sometimes invisible in blue, s-process elements enhanced over solar abundance, weak isotopic bands	thin disc pop I	-2.2	AGB	3100-2600	R Lep	~90

non-classical carbon stars							
C-J:	very strong isotopic bands of C_2 and CN	*unknown*	*unknown*	*unknown*	3900-2800	Y CVn	~20
C-H:	very strong CH absorption	halo pop II	-1.8	bright giants, mass transfer (all C-H:s are binary)	5000-4100	*V Ari, TT CVn*	~20
C-Hd:	hydrogen lines and CH bands weak or absent	thin disc pop I	-3.5	*unknown*	?	*HD 137613*	~7

Astrophysical Mechanisms

Carbon stars can be explained by more than one astrophysical mechanism. *Classical carbon stars* are distinguished from *non-classical* ones on the grounds of mass, with classical carbon stars being the more massive.

In the *classical carbon stars*, those belonging to the modern spectral types C-R and C-N, the abundance of carbon is thought to be a product of helium fusion, specifically the triple-alpha process within a star, which giants reach near the end of their lives in the asymptotic giant branch (AGB). These fusion products have been brought to the stellar surface by episodes of convection (the so-called third dredge-up) after the carbon and other products were made. Normally this kind of AGB carbon star fuses hydrogen in a hydrogen burning shell, but in episodes separated by 10^4-10^5 years, the star transforms to burning helium in a shell, while the hydrogen fusion temporarily ceases. In this phase, the star's luminosity rises, and material from the interior of the star (notably carbon) moves up. Since the luminosity rises, the star expands so that the helium fusion ceases, and the hydrogen shell burning restarts. During these *shell helium flashes*, the mass loss from the star is significant, and after many shell helium flashes, an AGB star is transformed into a hot white dwarf and its atmosphere becomes material for a planetary nebula.

The *non-classical* kinds of carbon stars, belonging to the types C-J and C-H, are believed to be binary stars, where one star is observed to be a giant star (or occasionally a red dwarf) and the other a white dwarf. The star presently observed to be a giant star accreted carbon-rich material when it was still a main-sequence star from its companion (that is, the star that is now the white dwarf) when the latter was still a classical carbon star. That phase of stellar evolution is relatively brief, and most such stars ultimately end up as white dwarfs. We are now seeing these systems a comparatively long time after the mass transfer event, so the extra carbon observed in the present red giant was not produced within that star. This scenario is also accepted as the origin of the barium stars, which are also characterized as having strong spectral features of carbon molecules and of barium (an s-process element). Sometimes the stars whose excess carbon came from this mass transfer are called "extrinsic" carbon stars to distinguish them from the "intrinsic" AGB stars which produce the carbon internally. Many of these extrinsic carbon stars are not luminous or cool enough to have made their own carbon, which was a puzzle until their binary nature was discovered.

The enigmatic *hydrogen deficient carbon stars* (HdC), belonging to the spectral class C-Hd, seems to have some relation to R Coronae Borealis variables (RCB), but are not variable themselves and lack a certain infrared radiation typical for RCB:s. Only five HdC:s are known, and none is known to be binary, so the relation to the non-classical carbon stars is not known.

Other less convincing theories, such as CNO cycle unbalancing and core helium flash have also been proposed as mechanisms for carbon enrichment in the atmospheres of smaller carbon stars.

Other Qualities

Most classical carbon stars are variable stars of the long period variable types.

Observing Carbon Stars

Due to the insensitivity of night vision to red and a slow adaption of the red sensitive eye rods to the light of the stars, astronomers making magnitude estimates of red variable stars, especially carbon stars, have to know how to deal with the Purkinje effect in order not to underestimate the magnitude of the observed star.

Interstellar Carbon sowers

Owing to its low surface gravity, as much as half (or more) of the total mass of a carbon star may be lost by way of powerful stellar winds. The star's remnants, carbon-rich "dust" similar to graphite, therefore become part of the interstellar dust. This dust is believed to be a significant factor in providing the raw materials for the creation of subsequent generations of stars and their planetary systems. The material surrounding a carbon star may blanket it to the extent that the dust absorbs all visible light.

Other Classifications

Other types of carbon stars include:

- CCS – Cool Carbon Star
- CEMP – Carbon-Enhanced Metal-Poor
  - CEMP-no – Carbon-Enhanced Metal-Poor star with no enhancement of elements produced by the r-process or s-process nucleosynthesis
  - CEMP-r – Carbon-Enhanced Metal-Poor star with an enhancement of elements produced by r-process nucleosynthesis
  - CEMP-s – Carbon-Enhanced Metal-Poor star with an enhancement of elements produced by s-process nucleosynthesis
  - CEMP-r/s – Carbon-Enhanced Metal-Poor star with an enhancement of elements produced by both r-process and s-process nucleosynthesis
- CGCS – Cool Galactic Carbon Star

Neutron Star

Radiation from the pulsar PSR B1509-58, a rapidly spinning neutron star, makes nearby gas glow in X-rays (gold, from Chandra) and illuminates the rest of the nebula, here seen in infrared (blue and red, from WISE).

A neutron star is the collapsed core of a large star (10–29 solar masses). Neutron stars are the smallest and densest stars known to exist. With a radius of only about 11–11.5 km (7 miles), they can, however, have a mass of about twice that of the Sun. They result from the supernova explosion of a massive star, combined with gravitational collapse, that compresses the core past the white dwarf star density to that of neutrons. Neutron stars are composed almost entirely of neutrons, which are subatomic particles with no net electrical charge and with slightly larger mass than protons. They are supported against further collapse by neutron degeneracy pressure, a phenomenon described by the Pauli exclusion principle. If the remnant has too great a mass, between 1.4 and 2-3 solar masses, it will continue to collapse into a form called a black hole. Neutron stars are very hot and typically have a surface temperature around 6×10^5 K. They are so dense that a normal-sized matchbox containing neutron-star material would have a mass of approximately 13 million tonnes, or a 2.5 million m^3 chunk of the Earth. The density of the star is comparable to that of the nucleus of an atom. They have strong magnetic fields, between 10^8 and 10^{15} times that of Earth's. The gravitational field at the neutron star's surface is about 2×10^{11} times that of the Earth's.

As the star's core collapses, its rotation rate increases as a result of conservation of angular momentum, hence neutron stars rotate at up to several hundred times per second. As they do so, they can emit beams of electromagnetic radiation that makes them detectable as pulsars. Indeed, the discovery of pulsars in 1967 first suggested that neutron stars exist. The radiation from pulsars is thought to be primarily emitted from regions near their magnetic poles. If the magnetic poles do not coincide with the rotational axis of the neutron star, the emission beam will sweep the sky, and when seen from a distance, if the observer is somewhere in the path of the beam, it will appear as pulses of radiation coming from a fixed point in space. The rotation of neutron stars can be very rapid; up to 716 times a second has been detected, which is approximately 43,000 revolutions per minute, giving a linear speed at the surface on the order of 0.165 c.

There are thought to be around 100 million neutron stars in the Milky Way, a figure obtained by estimating the number of stars that have undergone supernova explosions. However, most are old and cold, and neutron stars can only be easily detected in certain instances, such as if they are a pulsar or part of a binary system. Non-rotating and non-accreting neutron stars are virtually undetectable; however, the *Hubble Space Telescope* has observed one thermally radiating neutron star, called RX J185635-3754. The sudden collapse of rapidly-rotating high-mass stars, or the merger of binary neutron stars, may be the source of gamma-ray bursts. Soft gamma repeaters are conjectured to be a type of neutron star with very strong magnetic fields, known as magnetars, or alternatively, neutron stars with fossil disks around them.

Formation

Any main-sequence star with an initial mass of above 8 $M_\odot$ has the potential to produce a neutron star. As the star evolves away from the main sequence, subsequent nuclear burning produces an iron-rich core. When all nuclear fuel in the core has been exhausted, the core must be supported by degeneracy pressure alone. Further deposits of mass from shell burning cause the core to exceed the Chandrasekhar limit. Electron-degeneracy pressure is overcome and the core collapses further, sending temperatures soaring to over 70095000000000000000$\spadesuit$5×10^9 K. At these temperatures, photodisintegration (the breaking up of iron nuclei into alpha particles by high-energy gamma rays) occurs. As the temperature climbs even higher, electrons and protons combine to form neutrons via electron capture, releasing a flood of neutrinos. When densities reach nuclear density of 7017400000000000000$\spadesuit$4×10^{17} kg/m³, neutron degeneracy pressure halts the contraction. The infalling outer atmosphere of the star is halted and flung outwards by a flux of neutrinos produced in the creation of the neutrons, becoming a Type II or Type Ib supernova. The remnant left is a neutron star. If the remnant has a mass greater than about 5 $M_\odot$, it collapses further to become a black hole.

As the core of a massive star is compressed during a Type II supernova, Type Ib or Type Ic supernova, and collapses into a neutron star, it retains most of its angular momentum. But, because it has only a tiny fraction of its parent's radius (and therefore its moment of inertia is sharply reduced), a neutron star is formed with very high rotation speed, and then over a very long period it slows. Neutron stars are known that have rotation periods from about 1.4 ms to 30 s. The neutron star's density also gives it very high surface gravity, with typical values ranging from 10^{12} to 10^{13} m/s² (more than 10^{11} times that of Earth). One measure of such immense gravity is the fact that neutron stars have an escape velocity ranging from 100,000 km/s to 150,000 km/s, that is, from a third to half the speed of light. The neutron star's gravity accelerates infalling matter to tremendous speed. The force of its impact would likely destroy the object's component atoms, rendering all the matter identical, in most respects, to the rest of the neutron star.

Properties

Mass and Temperature

A neutron star has a mass of at least 1.1 and perhaps up to 3 solar masses ($M_\odot$). The maximum observed mass of neutron stars is about 2.01 $M_\odot$. But in general, compact stars of less than 1.39 $M_\odot$ (the Chandrasekhar limit) are white dwarfs, whereas compact stars with a mass between 1.4 $M_\odot$ and 3 $M_\odot$ (the Tolman–Oppenheimer–Volkoff limit) should be neutron stars. Between 3 $M_\odot$ and 5 $M_\odot$, hypothetical intermediate-mass stars such as quark stars and electroweak stars have been proposed, but none have been shown to exist. Beyond 10 $M_\odot$ the stellar remnant will overcome the neutron degeneracy pressure and gravitational collapse will usually occur to produce a black hole, though the smallest observed mass of a stellar black hole is about 5 $M_\odot$.

The temperature inside a newly formed neutron star is from around 10^{11} to 10^{12} kelvin. However, the huge number of neutrinos it emits carry away so much energy that the temperature falls within a few years to around 10^6 kelvin. Even at 1 million kelvin, most of the light generated by a neutron star is in X-rays.

Density and Pressure

Neutron stars have overall densities of 3.7×10^{17} to 5.9×10^{17} kg/m³ 2.6×10^{14} to 4.1×10^{14} times the density of the Sun),[b] which is comparable to the approximate density of an atomic nucleus of 3×10^{17} kg/m³. The neutron star's density varies from about 1×10^9 kg/m³ in the crust—increasing with depth—to about 6×10^{17} or 8×10^{17} kg/m³ (denser than an atomic nucleus) deeper inside. A neutron star is so dense that one teaspoon (5 milliliters) of its material would have a mass over 5.5×10^{12} kg (that is 1100 tonnes per 1 nanolitre), about 900 times the mass of the Great Pyramid of Giza.

The equations of state of matter at such high densities are not precisely known because of the the-oretical and empirical difficulties.

The pressure increases from 3×10^{33} to 1.6×10^{35} Pa from the inner crust to the center.

Giant Nucleus

A neutron star has some of the properties of an atomic nucleus, including density (within an order of magnitude) and being composed of nucleons. In popular scientific writing, neutron stars are therefore sometimes described as giant nuclei. However, in other respects, neutron stars and atomic nuclei are quite different. In particular, a nucleus is held together by the strong interaction, whereas a neutron star is held together by gravity, and thus the density and structure of neutron stars is more variable. It is generally more useful to consider such objects as stars.

Magnetic Field

Neutron stars have strong magnetic fields. The magnetic field strength on the surface of neutron stars have been estimated at least to have the range of 10^8 to 10^{15} gauss (10^4 to 10^{11} tesla). In comparison, the magnitude at Earth's surface ranges from 25 to 65 microteslas (0.25 to 0.65 gauss), making the field at least 10^8 times as strong as that of Earth. Variations in magnetic field strengths are most likely the main factor that allows different types of neutron stars to be distinguished by their spectra, and explains the periodicity of pulsars. The neutron stars known as magnetars have the strongest magnetic fields, in the range of 10^8 to 10^{11} tesla, and have become the widely accepted hypothesis for neutron star types soft gamma repeaters (SGRs) and anomalous X-ray pulsars (AXPs).

The origins of the strong magnetic field are as yet unclear. One hypothesis is that of "flux freezing", or conservation of the original magnetic flux takes place during the formation of the neutron star. If an object has a certain magnetic flux over its surface area, and that area shrinks to a smaller area, but the magnetic flux is conserved, then the magnetic field would correspondingly increase. Likewise, a collapsing star begins with a much larger surface area than the resulting neutron star, and conservation of magnetic flux would result in a far stronger magnetic field. However, this simple explanation does not fully explain magnetic field strengths of neutron stars.

Gravity and Equation of State

Gravitational light deflection at a neutron star. Due to relativistic light deflection more than half of the surface is visible (each chequered patch here represents 30 degrees by 30 degrees). In natural units, the mass of the depicted star is 1 and its radius 4, or twice its Schwarzschild radius.

The gravitational field at a neutron star's surface is about 2×10^{11} times stronger than on Earth, at around 1.86×10^8 m/s². Such a strong gravitational field acts as a gravitational lens and bends the radiation emitted by the neutron star such that parts of the normally invisible rear surface become visible. If the radius of the neutron star is $3GM/c^2$ or less, then the photons may be trapped in an orbit, thus making the whole surface of that neutron star visible, along with destabilizing orbits at that and less than that of the radius. A fraction of the mass of a star that collapses to form a neutron star is released in the supernova explosion from which it forms (from the law of mass-energy equivalence, $E = mc^2$). The energy comes from the gravitational binding energy of a neutron star.

Hence, the gravitational force of a typical neutron star is huge. If an object were to fall from a height of one meter in a neutron star 12 kilometers in radius, it would reach the ground at around 1.4 million meters per second, or 5 million kilometers per hour.

Because of the enormous gravity, time dilation between a neutron star and Earth is significant. For example, eight years could pass on the surface of a neutron star, yet ten years would have passed on Earth, not including the time-dilation effect of their very rapid rotation.

Neutron star relativistic equations of state describe the relation of radius vs. mass for various models. The most likely radii for a given neutron star mass are bracketed by models AP4 (smallest radius) and MS2 (largest radius). BE is the ratio of gravitational binding energy mass equivalent to the observed neutron star gravitational mass of "M" kilograms with radius "R" meters,

$$BE \quad \frac{0.60}{\underline{}\ _{000\ 00c}} \qquad \beta = GM/Rc^2$$

Given current values

$$G = 6.6742 \times 10^{-11} \text{m}^3 \text{kg}^{-1} \text{s}^{-2}$$

$$c^2 = 8.98755 \times 10^{16} \text{m}^2 \text{s}^{-2}$$

$$M_{solar} = 1.98844 \times 10^{30} \text{kg}$$

and star masses "M" commonly reported as multiples of one solar mass,

$$M_x = \frac{M}{M_\odot}$$

then the relativistic fractional binding energy of a neutron star is

$$BE = \frac{885.975 M_x}{R - 738.313 M_x}$$

A 2 $M_\odot$ neutron star would not be more compact than 10,970 meters radius (AP4 model). Its mass fraction gravitational binding energy would then be 0.187, −18.7% (exothermic). This is not near 0.6/2 = 0.3, −30%.

The equation of state for a neutron star is still not known. It is assumed that it differs significantly from that of a white dwarf, whose equation of state is that of a degenerate gas that can be described in close agreement with special relativity. However, with a neutron star the increased effects of general relativity can no longer be ignored. Several equations of state have been proposed (FPS, UU, APR, L, SLy, and others) and current research is still attempting to constrain the theories to make predictions of neutron star matter. This means that the relation between density and mass is not fully known, and this causes uncertainties in radius estimates. For example, a 1.5 $M_\odot$ neutron star could have a radius of 10.7, 11.1, 12.1 or 15.1 kilometers (for EOS FPS, UU, APR or L respectively).

Structure

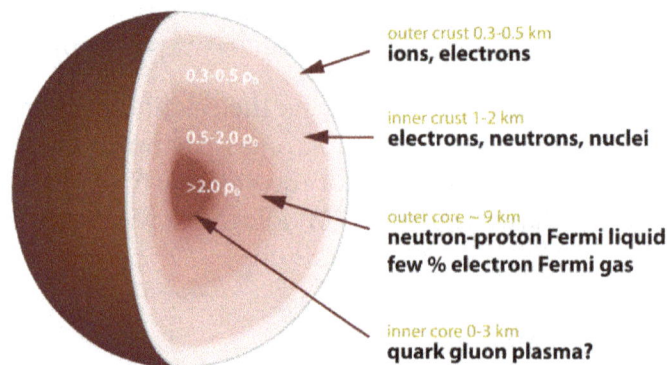

outer crust 0.3-0.5 km
ions, electrons

inner crust 1-2 km
electrons, neutrons, nuclei

outer core ~ 9 km
**neutron-proton Fermi liquid
few % electron Fermi gas**

inner core 0-3 km
quark gluon plasma?

0.3-0.5 ρ_0

0.5-2.0 ρ_0

>2.0 ρ_0

Cross-section of neutron star. Densities are in terms of ρ_o the saturation nuclear matter density, where nucleons begin to touch.

Current understanding of the structure of neutron stars is defined by existing mathematical models, but it might be possible to infer some details through studies of neutron-star oscillations. Asteroseismology, a study applied to ordinary stars, can reveal the inner structure of neutron stars by analyzing observed spectra of stellar oscillations.

Current models indicate that matter at the surface of a neutron star is composed of ordinary atomic nuclei crushed into a solid lattice with a sea of electrons flowing through the gaps between them. It is possible that the nuclei at the surface are iron, due to iron's high binding energy per nucleon. It is also possible that heavy elements, such as iron, simply sink beneath the surface, leaving only light nuclei like helium and hydrogen. If the surface temperature exceeds 10^6 kelvin (as in the case of a young pulsar), the surface should be fluid instead of the solid phase that might exist in cooler neutron stars (temperature <10^6 kelvin).

The "atmosphere" of a neutron star is hypothesized to be at most several micrometers thick, and its dynamics are fully controlled by the neutron star's magnetic field. Below the atmosphere one encounters a solid "crust". This crust is extremely hard and very smooth (with maximum surface irregularities of ~5 mm), due to the extreme gravitational field. The expected hierarchy of phases of nuclear matter in the inner crust has been characterized as nuclear pasta.

Proceeding inward, one encounters nuclei with ever increasing numbers of neutrons; such nuclei would decay quickly on Earth, but are kept stable by tremendous pressures. As this process continues at increasing depths, the neutron drip becomes overwhelming, and the concentration of free neutrons increases rapidly. In that region, there are nuclei, free electrons, and free neutrons. The nuclei become increasingly small (gravity and pressure overwhelming the strong force) until the core is reached, by definition the point where mostly neutrons exist.

The composition of the superdense matter in the core remains uncertain. One model describes the core as superfluid neutron-degenerate matter (mostly neutrons, with some protons and electrons). More exotic forms of matter are possible, including degenerate strange matter (containing strange quarks in addition to up and down quarks), matter containing high-energy pions and kaons in addition to neutrons, or ultra-dense quark-degenerate matter.

Radiation

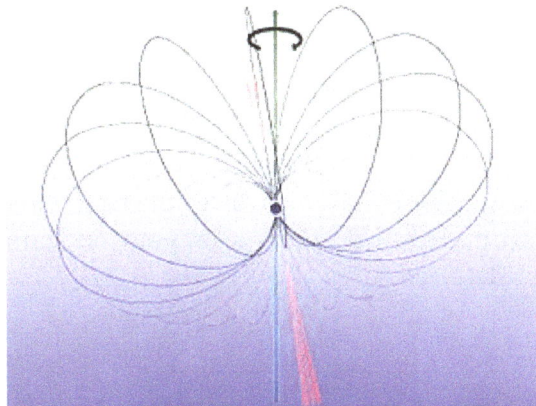

Animation of a rotating pulsar. The sphere in the middle represents the neutron star, the curves indicate the magnetic field lines and the protruding cones represent the emission zones.

Pulsars

Neutron stars are detected from their electromagnetic radiation. Neutron stars are usually observed to pulse radio waves and other electromagnetic radiation, and neutron stars observed with pulses are called pulsars.

Pulsars' radiation is thought to be caused by particle acceleration near their magnetic poles, which need not be aligned with the rotational axis of the neutron star. It is thought that a large electrostatic field builds up near the magnetic poles, leading to electron emission. These electrons are magnetically accelerated along the field lines, leading to curvature radiation, with the radiation being strongly polarized towards the plane of curvature. In addition, high energy photons can interact with lower energy photons and the magnetic field for electron-positron pair production, which through electron–positron annihilation leads to further high energy photons.

The radiation emanating from the magnetic poles of neutron stars can be described as *magnetospheric radiation*, in reference to the magnetosphere of the neutron star. It is not to be confused with *magnetic dipole radiation*, which is emitted because the magnetic axis is not aligned with the rotational axis, with a radiation frequency the same as the neutron star's rotational frequency.

If the axis of rotation of the neutron star is different to the magnetic axis, external viewers will only see these beams of radiation whenever the magnetic axis point towards them during the neutron star rotation. Therefore, periodic pulses are observed, at the same rate as the rotation of the neutron star.

Non-Pulsating Neutron stars

In addition to pulsars, neutron stars have also been identified with no apparent periodicity of their radiation. This seems to be a characteristic of the X-ray sources known as Central Compact Objects in Supernova remnants (CCOs in SNRs), which are thought to be young, radio-quiet isolated neutron stars.

Spectra

In addition to radio emissions, neutron stars have also been identified in other parts of the electromagnetic spectrum. This includes visible light, near infrared, ultraviolet X-rays and gamma rays. Pulsars observed in X-rays are known as X-ray pulsars if accretion-powered; while those identified in visible light as optical pulsars. The majority of neutron stars detected, including those identified in optical, X-ray and gamma rays, also emit radio waves; the Crab Pulsar produces electromagnetic emissions across the spectrum. However, there exist neutron stars called radio-quiet neutron stars, with no radio emissions detected.

Rotation

Neutron stars rotate extremely rapidly after their formation due to the conservation of angular momentum; like spinning ice skaters pulling in their arms, the slow rotation of the original star's core speeds up as it shrinks. A newborn neutron star can rotate many times a second.

Spin Down

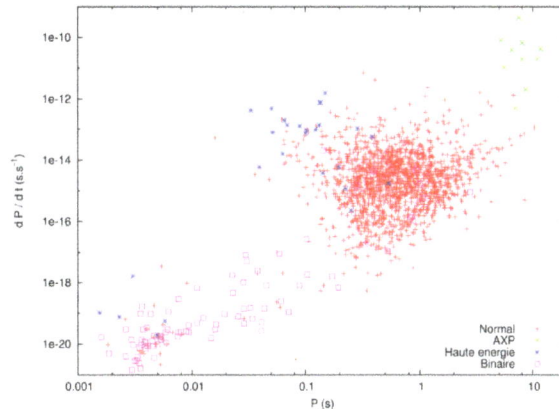

P–Pdot diagram for known rotation-powered pulsars (red), anomalous X-ray pulsars (green), high-energy emission pulsars (blue) and binary pulsars (pink)

Over time, neutron stars slow, as their rotating magnetic fields in effect radiate energy associated with the rotation; older neutron stars may take several seconds for each revolution. This is called *spin down*. The rate at which a neutron star slows its rotation is usually constant and very small.

The periodic time (P) is the rotational period, the time for one rotation of a neutron star. The spin-down rate, the rate of slowing of rotation, is then given the symbol $\dot{P}$ (or Pdot), the negative derivative of P with respect to time. It is defined as periodic time decrease per unit time; it is a dimensionless quantity, but can be given the units of s·s⁻¹ (seconds per second).

The spin-down rate (Pdot) of neutron stars usually falls within the range of 10^{-22} to 10^{-9} s·s⁻¹, with the shorter period (or faster rotating) observable neutron stars usually having smaller Pdot. However, as a neutron star ages, the neutron star slows (P increases) and the rate of slowing decreases (Pdot decreases), until the rate of rotation becomes too slow to power the radio-emission mechanism, and the neutron star can no longer be detected.

P and Pdot allow minimum magnetic fields of neutron stars to be estimated. P and Pdot can be also used to calculate the *characteristic age* of a pulsar, but gives an estimate which is somewhat larger than the true age when it is applied to young pulsars.

P and Pdot can also be combined with neutron star's moment of inertia to estimate a quantity called *spin-down luminosity*, which is given the symbol $\dot{E}$ (or Edot). It is not the measured luminosity, but rather the calculated loss rate of rotational energy that would manifest itself as radiation. For neutron stars where the spin-down luminosity is comparable to the actual luminosity, the neutron stars are said to be "rotation powered". The observed luminosity of the Crab Pulsar is comparable to the spin-down luminosity, supporting the model that rotational kinetic energy powers the radiation from it. With neutron stars such as magnetars, where the actual luminosity exceeds the spin-down luminosity by about a factor of one hundred, it is assumed that the luminosity is powered by magnetic dissipation, rather than being rotation powered.

P and Pdot can also be plotted for neutron stars to create a P-Pdot diagram. It encodes a tremendous amount of information about the pulsar population and its properties, and has been likened to the Hertzsprung–Russell diagram in its importance for neutron stars.

Spin Up

Neutron star rotational speeds can increase, a process known as spin up. Sometimes neutron stars absorbs orbiting matter from companion stars, increasing the rotation rate and reshaping the neutron star into an oblate spheroid. This causes an increase in the rate of rotation of the neutron star of over a hundred times per second in the case of millisecond pulsars.

The most rapidly rotating neutron star currently known, PSR J1748-2446ad, rotates at 716 rotations per second. However, a recent paper reported the detection of an X-ray burst oscillation, which provides an indirect measure of spin, of 1122 Hz from the neutron star XTE J1739-285, suggesting 1122 rotations a second. However, at present, this signal has only been seen once, and should be regarded as tentative until confirmed in another burst from that star.

Glitches and Starquakes

Sometimes a neutron star will undergo a glitch, a sudden small increase of its rotational speed or spin up. Glitches are thought to be the effect of a starquake—as the rotation of the neutron star slows, its shape becomes more spherical. Due to the stiffness of the "neutron" crust, this happens as discrete events when the crust ruptures, creating a starquake similar to earthquakes. After the starquake, the star will have a smaller equatorial radius, and because angular momentum is conserved, its rotational speed has increased.

Starquakes occurring in magnetars, with a resulting glitch, is the leading hypothesis for the gamma-ray sources known as soft gamma repeaters.

NASA artist's conception of a "starquake", or "stellar quake".

Recent work, however, suggests that a starquake would not release sufficient energy for a neutron star glitch; it has been suggested that glitches may instead be caused by transitions of vortices in the theoretical superfluid core of the neutron star from one metastable energy state to a lower one, thereby releasing energy that appears as an increase in the rotation rate.

"Anti-Glitches"

An "anti-glitch", a sudden small decrease in rotational speed, or spin down, of a neutron star has also been reported. It occurred in a magnetar, that in one case produced an X-ray luminosity increase of a factor of 20, and a significant spin-down rate change. Current neutron star models do not predict this behavior. If the cause was internal, it suggests differential rotation of solid outer crust and the superfluid component of the inner of the magnetar's structure.

Population and Distances

Central neutron star at the heart of the Crab Nebula.

At present, there are about 2000 known neutron stars in the Milky Way and the Magellanic Clouds, the majority of which have been detected as radio pulsars. Neutron stars are mostly concentrated along the disk of the Milky Way although the spread perpendicular to the disk is large because the supernova explosion process can impart high translational speeds (400 km/s) to the newly formed neutron star.

Some of the closest known neutron stars are RX J1856.5-3754, which is about 400 light years away, and PSR J0108-1431 at about 424 light years. RX J1856.5-3754 is a member of a close group of neutron stars called The Magnificent Seven. Another nearby neutron star that was detected transiting the backdrop of the constellation Ursa Minor has been nicknamed Calvera by its Canadian and American discoverers, after the villain in the 1960 film *The Magnificent Seven*. This rapidly moving object was discovered using the ROSAT/Bright Source Catalog.

Binary Neutron Star Systems

About 5% of all known neutron stars are members of a binary system. The formation and evolution of binary neutron stars can be a complex process. Neutron stars have been observed in binaries with ordinary main-sequence stars, red giants, white dwarfs or other neutron stars. According to modern theories of binary evolution it is expected that neutron stars also exist in binary systems with black hole companions. The merger of binaries containing two neutron stars, or a neutron star and a black hole, are expected to be prime sources for the emission of detectable gravitational waves.

Circinus X-1: X-ray light rings from a binary neutron star (24 June 2015; Chandra X-ray Observatory)

X-ray Binaries

Binary systems containing neutron stars often emit X-rays, which are emitted by hot gas as it falls towards the surface of the neutron star. The source of the gas is the companion star, the outer layers of which can be stripped off by the gravitational force of the neutron star if the two stars are sufficiently close. As the neutron star accretes this gas its mass can increase; if enough mass is accreted the neutron star may collapse into a black hole.

Neutron Star Binary Mergers and Nucleosynthesis

Binaries containing two neutron stars are observed to shrink as gravitational waves are emitted. Ultimately the neutron stars will come into contact and coelesce. The coalescence of binary neutron stars is one of the leading models for the origin of short gamma-ray bursts. Strong evidence for this model came from the observation of a kilonova associated with the short-duration gamma-ray burst GRB 130603B. The light emitted in the kilonova is believed to come from the radioactive decay of material ejected in the merger of the two neutron stars. This material may be responsible for the production of many of the chemical elements beyond iron, as opposed to the supernova nucleosynthesis theory.

History of Discoveries

In 1934, Walter Baade and Fritz Zwicky proposed the existence of neutron stars, only a year after the discovery of the neutron by Sir James Chadwick. In seeking an explanation for the origin of a supernova, they tentatively proposed that in supernova explosions ordinary stars are turned into stars that consist of extremely closely packed neutrons that they called neutron stars. Baade and Zwicky correctly proposed at that time that the release of the gravitational binding energy of the neutron stars powers the supernova: "In the supernova process, mass in bulk is annihilated". Neutron stars were thought to be too faint to be detectable and little work was done on them until November 1967, when Franco Pacini (1939–2012) pointed out that if the neutron stars were spinning and had large magnetic fields, then electromagnetic waves would be emitted. Unbeknown to him, radio astronomer Antony Hewish and his research assistant Jocelyn Bell at Cambridge were shortly to detect radio pulses from stars that are now believed to be highly magnetized, rapidly spinning neutron stars, known as pulsars.

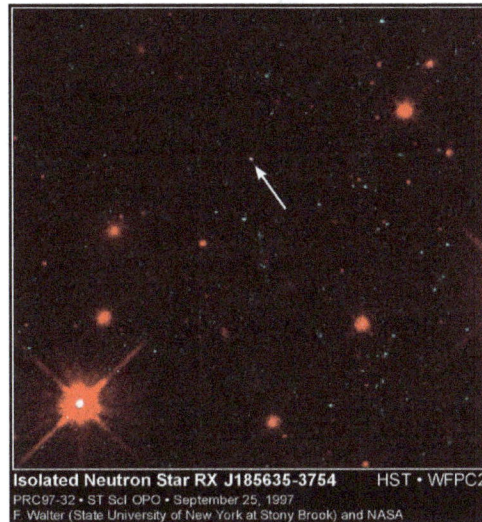

The first direct observation of a neutron star in visible light. The neutron star is RX J1856.5-3754.

In 1965, Antony Hewish and Samuel Okoye discovered "an unusual source of high radio brightness temperature in the Crab Nebula". This source turned out to be the Crab Pulsar that resulted from the great supernova of 1054.

In 1967, Iosif Shklovsky examined the X-ray and optical observations of Scorpius X-1 and correctly concluded that the radiation comes from a neutron star at the stage of accretion.

In 1967, Jocelyn Bell Burnell and Antony Hewish discovered regular radio pulses from PSR B1919+21. This pulsar was later interpreted as an isolated, rotating neutron star. The energy source of the pulsar is the rotational energy of the neutron star. The majority of known neutron stars (about 2000, as of 2010) have been discovered as pulsars, emitting regular radio pulses.

In 1971, Riccardo Giacconi, Herbert Gursky, Ed Kellogg, R. Levinson, E. Schreier, and H. Tananbaum discovered 4.8 second pulsations in an X-ray source in the constellation Centaurus, Cen X-3. They interpreted this as resulting from a rotating hot neutron star. The energy source is gravitational and results from a rain of gas falling onto the surface of the neutron star from a companion star or the interstellar medium.

In 1974, Antony Hewish was awarded the Nobel Prize in Physics "for his decisive role in the discovery of pulsars" without Jocelyn Bell who shared in the discovery.

In 1974, Joseph Taylor and Russell Hulse discovered the first binary pulsar, PSR B1913+16, which consists of two neutron stars (one seen as a pulsar) orbiting around their center of mass. Einstein's general theory of relativity predicts that massive objects in short binary orbits should emit gravitational waves, and thus that their orbit should decay with time. This was indeed observed, precisely as general relativity predicts, and in 1993, Taylor and Hulse were awarded the Nobel Prize in Physics for this discovery.

In 1982, Don Backer and colleagues discovered the first millisecond pulsar, PSR B1937+21. This objects spins 642 times per second, a value that placed fundamental constraints on the mass and radius of neutron stars. Many millisecond pulsars were later discovered, but PSR B1937+21 remained the fastest-spinning known pulsar for 24 years, until PSR J1748-2446ad (which spins more than 700 times a second) was discovered.

In 2003, Marta Burgay and colleagues discovered the first double neutron star system where both components are detectable as pulsars, PSR J0737-3039. The discovery of this system allows a total of 5 different tests of general relativity, some of these with unprecedented precision.

In 2010, Paul Demorest and colleagues measured the mass of the millisecond pulsar PSR J1614–2230 to be $1.97 \pm 0.04\ M_\odot$, using Shapiro delay. This was substantially higher than any previously measured neutron star mass, and places strong constraints on the interior composition of neutron stars.

In 2013, John Antoniadis and colleagues measured the mass of PSR J0348+0432 to be $2.01 \pm 0.04\ M_\odot$, using white dwarf spectroscopy. This confirmed the existence of such massive stars using a different method. Furthermore, this allowed, for the first time, a test of general relativity using such a massive neutron star.

Subtypes Table

- Neutron star
  - Isolated Neutron Star (INS): not in a binary system.
    - Rotation-powered pulsar (RPP or "radio pulsar"): neutron stars that emit directed pulses of radiation towards us at regular intervals (due to their strong magnetic fields).
      - Rotating radio transient (RRATs): are thought to be pulsars which emit more sporadically and/or with higher pulse-to-pulse variability than the bulk of the known pulsars.
    - Magnetar: a neutron star with an extremely strong magnetic field (1000 times more than a regular neutron star), and long rotation periods (5 to 12 seconds).
      - Soft gamma repeater (SGR).
      - Anomalous X-ray pulsar (AXP).
    - Radio-quiet neutron stars.
      - X-ray Dim Isolated Neutron Stars.
      - Central Compact Objects in Supernova remnants (CCOs in SNRs): young, radio-quiet non-pulsating X-ray sources, thought to be Isolated Neutron Stars surrounded by supernova remnants.
  - X-ray pulsars or "accretion-powered pulsars": a class of X-ray binaries.
    - Low-mass X-ray binary pulsars: a class of low-mass X-ray binaries (LMXB), a pulsar with a main sequence star, white dwarf or red giant.

- Millisecond pulsar (MSP) ("recycled pulsar").

  - Sub-millisecond pulsar.

- X-ray burster: a neutron star with a low mass binary companion from which matter is accreted resulting in irregular bursts of energy from the surface of the neutron star.

  - Intermediate-mass X-ray binary pulsars: a class of intermediate-mass X-ray binaries (IMXB), a pulsar with an intermediate mass star.

  - High-mass X-ray binary pulsars: a class of high-mass X-ray binaries (HMXB), a pulsar with a massive star.

  - Binary pulsars: a pulsar with a binary companion, often a white dwarf or neutron star.

- Theorized compact stars with similar properties.

  - Protoneutron star (PNS), theorized.

  - Exotic star

    - Quark star: currently a hypothetical type of neutron star composed of quark matter, or strange matter. As of 2008, there are three candidates.

    - Electroweak star: currently a hypothetical type of extremely heavy neutron star, in which the quarks are converted to leptons through the electroweak force, but the gravitational collapse of the neutron star is prevented by radiation pressure. As of 2010, there is no evidence for their existence.

    - Preon star: currently a hypothetical type of neutron star composed of preon matter. As of 2008, there is no evidence for the existence of preons.

Examples of Neutron stars

- PSR J0108-1431 – closest neutron star

- LGM-1 – the first recognized radio-pulsar

- PSR B1257+12 – the first neutron star discovered with planets (a millisecond pulsar)

- SWIFT J1756.9-2508 – a millisecond pulsar with a stellar-type companion with planetary range mass (below brown dwarf)

- PSR B1509-58 source of the "Hand of God" photo shot by the Chandra X-ray Observatory.

- PSR J0348+0432 – the most massive neutron star with a well-constrained mass, $2.01 \pm 0.04\, M_\odot$.

References

- North, John (1995). The Norton History of Astronomy and Cosmology. New York and London: W.W. Norton & Company. pp. 30–31. ISBN 0-393-03656-1.

- Murdin, P. (November 2000). "Aristillus (c. 200 BC)". Encyclopedia of Astronomy and Astrophysics. Bibcode:2000eaa..bookE3440.. doi:10.1888/0333750888/3440. ISBN 0-333-75088-8.

- Koch-Westenholz, Ulla; Koch, Ulla Susanne (1995). Mesopotamian astrology: an introduction to Babylonian and Assyrian celestial divination. Carsten Niebuhr Institute Publications. 19. Museum Tusculanum Press. p. 163. ISBN 87-7289-287-0.

- Lyall, Francis; Larsen, Paul B. (2009). "Chapter 7: The Moon and Other Celestial Bodies". Space Law: A Treatise. Ashgate Publishing, Ltd. p. 176. ISBN 0-7546-4390-5.

- Plait, Philip C. (2002). Bad astronomy: misconceptions and misuses revealed, from astrology to the moon landing "hoax". John Wiley and Sons. pp. 237–240. ISBN 0-471-40976-6.

- Kwok, Sun (2000). The origin and evolution of planetary nebulae. Cambridge astrophysics series. 33. Cambridge University Press. pp. 103–104. ISBN 0-521-62313-8.

- Szebehely, Victor G.; Curran, Richard B. (1985). Stability of the Solar System and Its Minor Natural and Artificial Bodies. Springer. ISBN 90-277-2046-0.

- Gray, David F. (1992). The Observation and Analysis of Stellar Photospheres. Cambridge University Press. pp. 413–414. ISBN 0-521-40868-7.

- Jørgensen, Uffe G. (1997), "Cool Star Models", in van Dishoeck, Ewine F., Molecules in Astrophysics: Probes and Processes, International Astronomical Union Symposia. Molecules in Astrophysics: Probes and Processes, 178, Springer Science & Business Media, p. 446, ISBN 079234538X

- Zeilik, Michael A.; Gregory, Stephan A. (1998). Introductory Astronomy & Astrophysics (4th ed.). Saunders College Publishing. p. 321. ISBN 0-03-006228-4.

- Jaschek, Carlos; Jaschek, Mercedes (1990). The Classification of Stars. Cambridge University Press. pp. 31–48. ISBN 0-521-38996-8.

- Prialnik, Dina (2000). An Introduction to the Theory of Stellar Structure and Evolution. Cambridge University Press. 195–212. ISBN 0-521-65065-8.

- Alves, J.; Lada, C.; Lada, E. (2001). Tracing H_2 Via Infrared Dust Extinction. Cambridge University Press. p. 217. ISBN 0-521-78224-4.

- Prialnik, Dina (2000). An Introduction to the Theory of Stellar Structure and Evolution. Cambridge University Press. pp. 198–199. ISBN 0-521-65937-X.

- Ballesteros-Paredes, J.; Klessen, R. S.; Mac Low, M.-M.; Vazquez-Semadeni, E. "Molecular Cloud Turbulence and Star Formation". In Reipurth, B.; Jewitt, D.; Keil, K. Protostars and Planets V. pp. 63–80. ISBN 0-8165-2654-0.

A Comprehensive Study of Black Holes

The major components of black holes are discussed in this chapter. The chapter strategically encompasses and incorporates the major components and key concepts of black holes, such as, black hole starship, black hole information paradox and black hole complementarity.

Black Hole

A black hole is a region of spacetime exhibiting such strong gravitational effects that nothing— not even particles and electromagnetiradiation such as light—can escape from inside it. The theory of general relativity predicts that a sufficiently compact mass can deform spacetime to form a black hole. The boundary of the region from which no escape is possible is called the event horizon. Although crossing the event horizon has enormous effect on the fate of the object crossing it, it appears to have no locally detectable features. In many ways a black hole acts like an ideal black body, as it reflects no light. Moreover, quantum field theory in curved spacetime predicts that event horizons emit Hawking radiation, with the same spectrum as a black body of a temperature inversely proportional to its mass. This temperature is on the order of billionths of a kelvin for black holes of stellar mass, making it essentially impossible to observe.

Objects whose gravitational fields are too strong for light to escape were first considered in the 18th century by John Michell and Pierre-Simon Laplace. The first modern solution of general relativity that would characterize a black hole was found by Karl Schwarzschild in 1916, although its interpretation as a region of space from which nothing can escape was first published by David Finkelstein in 1958. Black holes were long considered a mathematical curiosity; it was during the 1960s that theoretical work showed they were a generiprediction of general relativity. The discovery of neutron stars sparked interest in gravitationally collapsed compact objects as a possible astrophysical reality.

Black holes of stellar mass are expected to form when very massive stars collapse at the end of their life cycle. After a black hole has formed, it can continue to grow by absorbing mass from its surroundings. By absorbing other stars and merging with other black holes, supermassive black holes of millions of solar masses ($M_\odot$) may form. There is general consensus that supermassive black holes exist in the centers of most galaxies.

Despite its invisible interior, the presence of a black hole can be inferred through its interaction with other matter and with electromagnetiradiation such as visible light. Matter that falls onto a black hole can form an external accretion disk heated by friction, forming some of the brightest objects in the universe. If there are other stars orbiting a black hole, their orbits can be used to determine the black hole's mass and location. Such observations can be used to exclude possible alternatives such as neutron stars. In this way, astronomers have identified numerous stellar black hole candidates in binary systems, and established that the radio source known as Sagittarius A*, at the core of our own Milky Way galaxy, contains a supermassive black hole of about 4.3 million solar masses.

On 11 February 2016, the LIGO collaboration announced the first observation of gravitational waves; because these waves were generated from a black hole merger it was the first ever direct detection of a binary black hole merger. On 15 June 2016, a second detection of a gravitational wave event from colliding black holes was announced.

Simulation of gravitational lensing by a black hole, which distorts the image of a galaxy in the background

Predicted appearance of non-rotating black hole with toroidal ring of ionised matter, such as has been proposed as a model for Sagittarius A*. The asymmetry is due to the Doppler effect resulting from the enormous orbital speed needed for centrifugal balance of the very strong gravitational attraction of the hole.

History

Simulated view of a black hole in front of the Large MagellaniCloud. Note the gravitational lensing effect, which produces two enlarged but highly distorted views of the Cloud. Across the top, the Milky Way disk appears distorted into an arc.

If the semi-diameter of a sphere of the same density as the Sun were to exceed that of the Sun in the proportion of 500 to 1, a body falling from an infinite height towards it would have acquired at its surface greater velocity than that of light, and consequently supposing light to be attracted by the same force in proportion to its vis inertiae, with other bodies, all light emitted from such a body would be made to return towards it by its own proper gravity.

—John Michell

In 1796, mathematician Pierre-Simon Laplace promoted the same idea in the first and second editions of his book *Exposition du système du Monde* (it was removed from later editions). He justified his argument mathematically in 1799.

Such "dark stars" were largely ignored in the nineteenth century, since it was not understood how a massless wave such as light could be influenced by gravity.

General Relativity

In 1915, Albert Einstein developed his theory of general relativity, having earlier shown that gravity does influence light's motion. Only a few months later, Karl Schwarzschild found a solution to the Einstein field equations, which describes the gravitational field of a point mass and a spherical mass. A few months after Schwarzschild, Johannes Droste, a student of Hendrik Lorentz, independently gave the same solution for the point mass and wrote more extensively about its properties. This solution had a peculiar behaviour at what is now called the Schwarzschild radius, where it became singular, meaning that some of the terms in the Einstein equations became infinite. The nature of this surface was not quite understood at the time. In 1924, Arthur Eddington showed that the singularity disappeared after a change of coordinates, although it took until 1933 for Georges Lemaître to realize that this meant the singularity at the Schwarzschild radius was an unphysical coordinate singularity. Arthur Eddington did however comment on the possibility of a star with mass compressed to the Schwarzschild radius in a 1926 book, noting that Einstein's theory allows us to rule out overly large densities for visible stars like Betelgeuse because "a star of 250 million km radius could not possibly have so high a density as the sun. Firstly, the force of gravitation would be so great that light would be unable to escape from it, the rays falling back to the star like a stone to the earth. Secondly, the red shift of the spectral lines would be so great that the spectrum would be shifted out of existence. Thirdly, the mass would produce so much curvature of the space-time metrithat space would close up around the star, leaving us outside (i.e., nowhere)."

In 1931, Subrahmanyan Chandrasekhar calculated, using special relativity, that a non-rotating body of electron-degenerate matter above a certain limiting mass (now called the Chandrasekhar limit at 1.4 $M_\odot$) has no stable solutions. His arguments were opposed by many of his contemporaries like Eddington and Lev Landau, who argued that some yet unknown mechanism would stop the collapse. They were partly correct: a white dwarf slightly more massive than the Chandrasekhar limit will collapse into a neutron star, which is itself stable because of the Pauli exclusion principle. But in 1939, Robert Oppenheimer and others predicted that neutron stars above approximately 3 $M_\odot$ (the Tolman–Oppenheimer–Volkoff limit) would collapse into black holes for the reasons presented by Chandrasekhar, and concluded that no law of physics was likely to intervene and stop at least some stars from collapsing to black holes.

Oppenheimer and his co-authors interpreted the singularity at the boundary of the Schwarzschild radius as indicating that this was the boundary of a bubble in which time stopped. This is a valid point of view for external observers, but not for infalling observers. Because of this property, the collapsed stars were called "frozen stars", because an outside observer would see the surface of the star frozen in time at the instant where its collapse takes it inside the Schwarzschild radius.

Golden Age

In 1958, David Finkelstein identified the Schwarzschild surface as an event horizon, "a perfect unidirectional membrane: causal influences can cross it in only one direction". This did not strictly contradict Oppenheimer's results, but extended them to include the point of view of infalling observers. Finkelstein's solution extended the Schwarzschild solution for the future of observers falling into a black hole. A complete extension had already been found by Martin Kruskal, who was urged to publish it.

These results came at the beginning of the golden age of general relativity, which was marked by general relativity and black holes becoming mainstream subjects of research. This process was helped by the discovery of pulsars in 1967, which, by 1969, were shown to be rapidly rotating neutron stars. Until that time, neutron stars, like black holes, were regarded as just theoretical curiosities; but the discovery of pulsars showed their physical relevance and spurred a further interest in all types of compact objects that might be formed by gravitational collapse.

In this period more general black hole solutions were found. In 1963, Roy Kerr found the exact solution for a rotating black hole. Two years later, Ezra Newman found the axisymmetrisolution for a black hole that is both rotating and electrically charged. Through the work of Werner Israel, Brandon Carter, and David Robinson the no-hair theorem emerged, stating that a stationary black hole solution is completely described by the three parameters of the Kerr–Newman metric: mass, angular momentum, and electricharge.

At first, it was suspected that the strange features of the black hole solutions were pathological artifacts from the symmetry conditions imposed, and that the singularities would not appear in generisituations. This view was held in particular by Vladimir Belinsky, Isaak Khalatnikov, and Evgeny Lifshitz, who tried to prove that no singularities appear in generisolutions. However, in the late 1960s Roger Penrose and Stephen Hawking used global techniques to prove that singularities appear generically.

Work by James Bardeen, Jacob Bekenstein, Carter, and Hawking in the early 1970s led to the formulation of black hole thermodynamics. These laws describe the behaviour of a black hole in close analogy to the laws of thermodynamics by relating mass to energy, area to entropy, and surface gravity to temperature. The analogy was completed when Hawking, in 1974, showed that quantum field theory predicts that black holes should radiate like a black body with a temperature proportional to the surface gravity of the black hole.

The first use of the term "black hole" in print was by journalist Ann Ewing in her article *"Black Holes' in Space"*, dated 18 January 1964, which was a report on a meeting of the American Association for the Advancement of Science. John Wheeler used the term "black hole" at a lecture in 1967, leading some to credit him with coining the phrase. After Wheeler's use of the term, it was quickly adopted in general use.

Properties and Structure

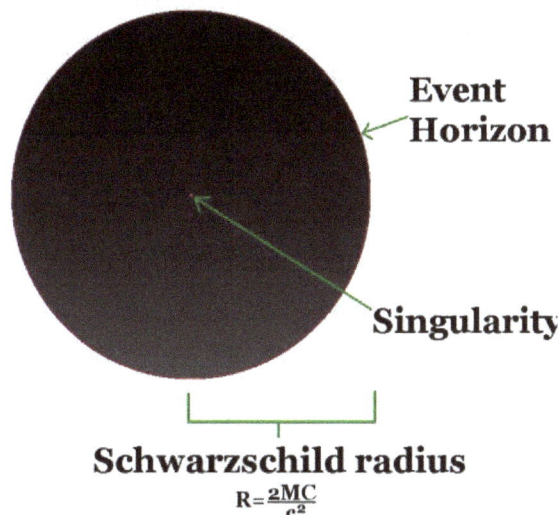

A simple illustration of a non-spinning black hole

The no-hair theorem states that, once it achieves a stable condition after formation, a black hole has only three independent physical properties: mass, charge, and angular momentum. Any two black holes that share the same values for these properties, or parameters, are indistinguishable according to classical (i.e. non-quantum) mechanics.

These properties are special because they are visible from outside a black hole. For example, a charged black hole repels other like charges just like any other charged object. Similarly, the total mass inside a sphere containing a black hole can be found by using the gravitational analog of Gauss's law, the ADM mass, far away from the black hole. Likewise, the angular momentum can be measured from far away using frame dragging by the gravitomagnetifield.

When an object falls into a black hole, any information about the shape of the object or distribution of charge on it is evenly distributed along the horizon of the black hole, and is lost to outside observers. The behavior of the horizon in this situation is a dissipative system that is closely analogous to that of a conductive stretchy membrane with friction and electrical resistance—the membrane paradigm. This is different from other field theories such as electromagnetism, which do not have any friction or resistivity at the microscopilevel, because they are time-reversible. Because a black hole eventually achieves a stable state with only three parameters, there is no way to avoid losing information about the initial conditions: the gravitational and electrifields of a black hole give very little information about what went in. The information that is lost includes every quantity that cannot be measured far away from the black hole horizon, including approximately conserved quantum numbers such as the total baryon number and lepton number. This behavior is so puzzling that it has been called the black hole information loss paradox.

Physical Properties

The simplest statiblack holes have mass but neither electricharge nor angular momentum. These black holes are often referred to as Schwarzschild black holes after Karl Schwarzschild who discovered this solution in 1916. According to Birkhoff's theorem, it is the only vacuum solution that is spherically symmetric. This means that there is no observable difference between the gravitational field of such a black hole and that of any other spherical object of the same mass. The popular notion of a black hole "sucking in everything" in its surroundings is therefore only correct near a black hole's horizon; far away, the external gravitational field is identical to that of any other body of the same mass.

Solutions describing more general black holes also exist. Non-rotating charged black holes are described by the Reissner–Nordström metric, while the Kerr metridescribes a non-charged rotating black hole. The most general stationary black hole solution known is the Kerr–Newman metric, which describes a black hole with both charge and angular momentum.

While the mass of a black hole can take any positive value, the charge and angular momentum are constrained by the mass. In Planck units, the total electricharge Q and the total angular momentum J are expected to satisfy

$$Q^2 + \left(\tfrac{J}{M}\right)^2 \le M^2$$

for a black hole of mass M. Black holes satisfying this inequality are called extremal. Solutions of Einstein's equations that violate this inequality exist, but they do not possess an event horizon. These solutions have so-called naked singularities that can be observed from the outside, and hence are deemed *unphysical*. The cosmicensorship hypothesis rules out the formation of such singularities, when they are created through the gravitational collapse of realistimatter. This is supported by numerical simulations.

Due to the relatively large strength of the electromagnetiforce, black holes forming from the collapse of stars are expected to retain the nearly neutral charge of the star. Rotation, however, is expected to be a common feature of compact objects. The black-hole candidate binary X-ray source GRS 1915+105 appears to have an angular momentum near the maximum allowed value.

Black hole classifications		
Class	Mass	Size
Supermassive black hole	~10^5–10^{10} M_{Sun}	~0.001–400 AU
Intermediate-mass black hole	~10^3 M_{Sun}	~10^3 km ≈ R_{Earth}
Stellar black hole	~10 M_{Sun}	~30 km
Micro black hole	up to ~M_{Moon}	up to ~0.1 mm

Black holes are commonly classified according to their mass, independent of angular momentum J or electricharge Q. The size of a black hole, as determined by the radius of the event horizon, or Schwarzschild radius, is roughly proportional to the mass M through

$$r_{sh} = \frac{2GM}{c^2} \approx 2.95 \frac{M}{M_{Sun}} \, \text{km},$$

where r_{sh} is the Schwarzschild radius and M_{Sun} is the mass of the Sun. This relation is exact only for black holes with zero charge and angular momentum; for more general black holes it can differ up to a factor of 2.

Event Horizon

Far away from the black hole, a particle can move in any direction, as illustrated by the set of arrows. It is only restricted by the speed of light.

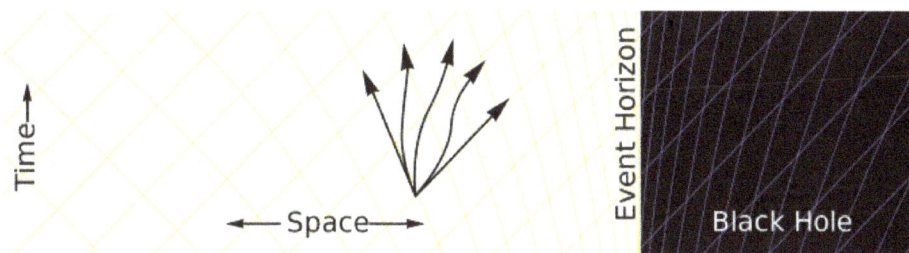

Closer to the black hole, spacetime starts to deform. There are more paths going towards the black hole than paths moving away.[Note 1]

Inside of the event horizon, all paths bring the particle closer to the center of the black hole. It is no longer possible for the particle to escape.

The defining feature of a black hole is the appearance of an event horizon—a boundary in spacetime through which matter and light can only pass inward towards the mass of the black hole. Nothing, not even light, can escape from inside the event horizon. The event horizon is referred to as such because if an event occurs within the boundary, information from that event cannot reach an outside observer, making it impossible to determine if such an event occurred.

As predicted by general relativity, the presence of a mass deforms spacetime in such a way that the paths taken by particles bend towards the mass. At the event horizon of a black hole, this deformation becomes so strong that there are no paths that lead away from the black hole.

To a distant observer, clocks near a black hole appear to tick more slowly than those further away from the black hole. Due to this effect, known as gravitational time dilation, an object falling into a black hole appears to slow as it approaches the event horizon, taking an infinite time to reach it. At the same time, all processes on this object slow down, from the view point of a fixed outside observer, causing any light emitted by the object to appear redder and dimmer, an effect known as gravitational redshift. Eventually, the falling object becomes so dim that it can no longer be seen.

On the other hand, indestructible observers falling into a black hole do not notice any of these effects as they cross the event horizon. According to their own clocks, which appear to them to tick normally, they cross the event horizon after a finite time without noting any singular behaviour; it is impossible to determine the location of the event horizon from local observations.

The shape of the event horizon of a black hole is always approximately spherical. For non-rotating (static) black holes the geometry of the event horizon is precisely spherical, while for rotating black holes the sphere is oblate.

Singularity

At the center of a black hole, as described by general relativity, lies a gravitational singularity, a region where the spacetime curvature becomes infinite. For a non-rotating black hole, this region takes the shape of a single point and for a rotating black hole, it is smeared out to form a ring singularity that lies in the plane of rotation. In both cases, the singular region has zero volume. It can also be shown that the singular region contains all the mass of the black hole solution. The singular region can thus be thought of as having infinite density.

Observers falling into a Schwarzschild black hole (*i.e.*, non-rotating and not charged) cannot avoid being carried into the singularity, once they cross the event horizon. They can prolong the experience by accelerating away to slow their descent, but only up to a limit; after attaining a certain ideal velocity, it is best to free fall the rest of the way. When they reach the singularity, they are crushed to infinite density and their mass is added to the total of the black hole. Before that happens, they will have been torn apart by the growing tidal forces in a process sometimes referred to as spaghettification or the "noodle effect".

In the case of a charged (Reissner–Nordström) or rotating (Kerr) black hole, it is possible to avoid the singularity. Extending these solutions as far as possible reveals the hypothetical possibility of exiting the black hole into a different spacetime with the black hole acting as a wormhole. The possibility of traveling to another universe is, however, only theoretical since any perturbation would destroy this possibility. It also appears to be possible to follow closed timelike curves (returning to one's own past) around the Kerr singularity, which lead to problems with causality like the grandfather paradox. It is expected that none of these peculiar effects would survive in a proper quantum treatment of rotating and charged black holes.

The appearance of singularities in general relativity is commonly perceived as signaling the breakdown of the theory. This breakdown, however, is expected; it occurs in a situation where quantum effects should describe these actions, due to the extremely high density and therefore particle interactions. To date, it has not been possible to combine quantum and gravitational effects into a single theory, although there exist attempts to formulate such a theory of quantum gravity. It is generally expected that such a theory will not feature any singularities.

Photon Sphere

The photon sphere is a spherical boundary of zero thickness in which photons that move on tangents to that sphere would be trapped in a circular orbit about the black hole. For non-rotating black holes, the photon sphere has a radius 1.5 times the Schwarzschild radius. Their orbits would be dynamically unstable, hence any small perturbation, such as a particle of infalling matter, would cause an instability that would grow over time, either setting the photon on an outward trajectory causing it to escape the black hole, or on an inward spiral where it would eventually cross the event horizon.

While light can still escape from the photon sphere, any light that crosses the photon sphere on an inbound trajectory will be captured by the black hole. Hence any light that reaches an outside observer from the photon sphere must have been emitted by objects between the photon sphere and the event horizon.

Other compact objects, such as neutron stars, can also have photon spheres. This follows from the fact that the gravitational field *external* to a spherically-symmetriobject is governed by the Schwarzschild metric, which depends only on the object's mass rather than the radius of the object, hence any object whose radius shrinks to smaller than 1.5 times the Schwarzschild radius will have a photon sphere.

Ergosphere

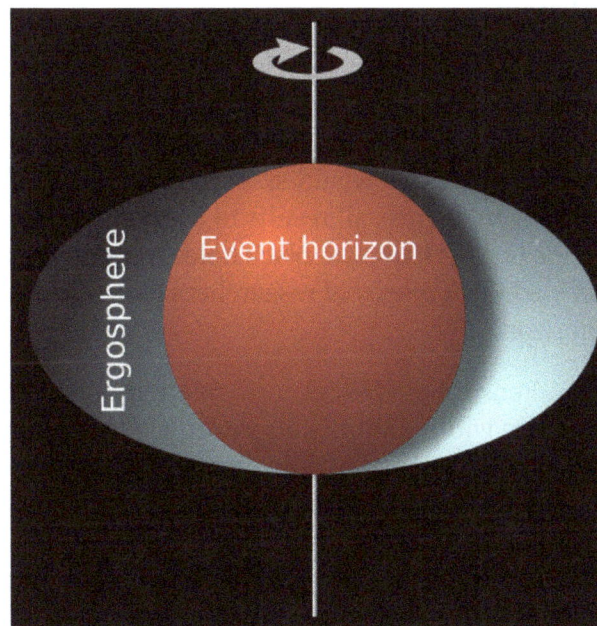

The ergosphere is an oblate spheroid region outside of the event horizon, where objects cannot remain stationary.

Rotating black holes are surrounded by a region of spacetime in which it is impossible to stand still, called the ergosphere. This is the result of a process known as frame-dragging; general relativity predicts that any rotating mass will tend to slightly "drag" along the spacetime immediately surrounding it. Any object near the rotating mass will tend to start moving in the direction of rotation. For a rotating black hole, this effect is so strong near the event horizon that an object would have to move faster than the speed of light in the opposite direction to just stand still.

The ergosphere of a black hole is a volume whose inner boundary is the black hole's event horizon and an outer boundary of an oblate spheroid, which coincides with the event horizon at the poles but noticeably wider around the equator. The outer boundary is sometimes called the *ergosurface.*

Objects and radiation can escape normally from the ergosphere. Through the Penrose process, objects can emerge from the ergosphere with more energy than they entered. This energy is taken from the rotational energy of the black hole causing the latter to slow.

Innermost Stable Circular Orbit (ISCO)

In Newtonian gravity, test particles can stably orbit at arbitrary distances from a central object. In general relativity, however, there exists an innermost stable circular orbit (often called the ISCO), inside of which, any infinitesimal perturbations to a circular orbit will lead to inspiral into the black hole. The location of the ISCO depends on the spin of the black hole, in the case of a Schwarzschild black hole (spin zero) is:

$$r_{isco} = 3r_s = \frac{6GM}{c^2},$$

and decreases with increasing spin.

Formation and Evolution

Considering the exotinature of black holes, it may be natural to question if such bizarre objects could exist in nature or to suggest that they are merely pathological solutions to Einstein's equations. Einstein himself wrongly thought that black holes would not form, because he held that the angular momentum of collapsing particles would stabilize their motion at some radius. This led the general relativity community to dismiss all results to the contrary for many years. However, a minority of relativists continued to contend that black holes were physical objects, and by the end of the 1960s, they had persuaded the majority of researchers in the field that there is no obstacle to the formation of an event horizon.

Once an event horizon forms, Penrose proved, general relativity without quantum mechanics requires that a singularity will form within. Shortly afterwards, Hawking showed that many cosmological solutions that describe the Big Bang have singularities without scalar fields or other exotimatter. The Kerr solution, the no-hair theorem, and the laws of black hole thermodynamics showed that the physical properties of black holes were simple and comprehensible, making them respectable subjects for research. The primary formation process for black holes is expected to be the gravitational collapse of heavy objects such as stars, but there are also more exotiprocesses that can lead to the production of black holes.

Gravitational Collapse

Gravitational collapse occurs when an object's internal pressure is insufficient to resist the object's own gravity. For stars this usually occurs either because a star has too little "fuel" left to maintain its temperature through stellar nucleosynthesis, or because a star that would have been stable receives extra matter in a way that does not raise its core temperature. In either case the star's temperature is no longer high enough to prevent it from collapsing under its own weight. The collapse may be stopped by the degeneracy pressure of the star's constituents, allowing the condensation of matter into an exotidenser state. The result is one of the various types of compact star. The type of compact star formed depends on the mass of the remnant of the original star left after the outer layers have been blown away. Such explosions and pulsations lead to planetary nebula. This mass can be substantially less than the original star. Remnants exceeding $5 \, M_\odot$ are produced by stars that were over $20 \, M_\odot$ before the collapse.

If the mass of the remnant exceeds about $3–4 \, M_\odot$ (the Tolman–Oppenheimer–Volkoff limit), either because the original star was very heavy or because the remnant collected additional mass through accretion of matter, even the degeneracy pressure of neutrons is insufficient to stop the collapse. No known mechanism is powerful enough to stop the implosion and the object will inevitably collapse to form a black hole.

Artist's impression of supermassive black hole seed.

The gravitational collapse of heavy stars is assumed to be responsible for the formation of stellar mass black holes. Star formation in the early universe may have resulted in very massive stars, which upon their collapse would have produced black holes of up to $10^3 \, M_\odot$. These black holes could be the seeds of the supermassive black holes found in the centers of most galaxies. It has further been suggested that supermassive black holes with typical masses of $\sim 10^5 \, M_\odot$ could have formed from the direct collapse of gas clouds in the young universe. Some candidates for such objects have been found in observations of the young universe.

While most of the energy released during gravitational collapse is emitted very quickly, an outside observer does not actually see the end of this process. Even though the collapse takes a finite amount of time from the reference frame of infalling matter, a distant observer would see the infalling material slow and halt just above the event horizon, due to gravitational time dilation. Light from the collapsing material takes longer and longer to reach the observer, with the light emitted just before the event horizon forms delayed an infinite amount of time. Thus the external observer never sees the formation of the event horizon; instead, the collapsing material seems to become dimmer and increasingly red-shifted, eventually fading away.

Primordial Black Holes in the Big Bang

Gravitational collapse requires great density. In the current epoch of the universe these high densities are only found in stars, but in the early universe shortly after the big bang densities were much greater, possibly allowing for the creation of black holes. The high density alone is not enough to allow the formation of black holes since a uniform mass distribution will not allow the mass to bunch up. In order for primordial black holes to form in such a dense medium, there must be initial density perturbations that can then grow under their own gravity. Different models for the early universe vary widely in their predictions of the size of these perturbations. Various models predict the creation of black holes, ranging from a Planck mass to hundreds of thousands of solar masses. Primordial black holes could thus account for the creation of any type of black hole.

High-Energy Collisions

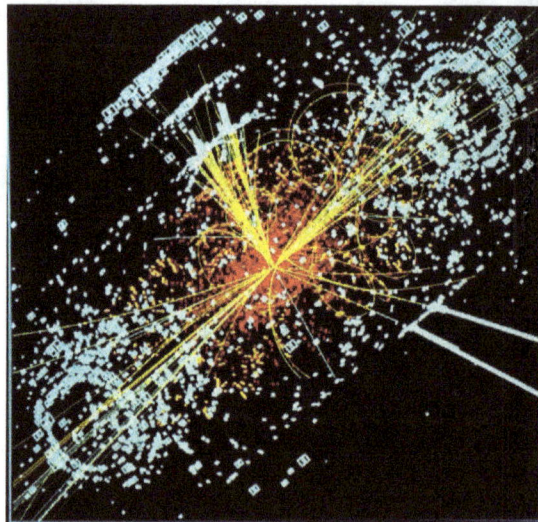

A simulated event in the CMS detector, a collision in which a micro black hole may be created.

Gravitational collapse is not the only process that could create black holes. In principle, black holes could be formed in high-energy collisions that achieve sufficient density. As of 2002, no such events have been detected, either directly or indirectly as a deficiency of the mass balance in particle accelerator experiments. This suggests that there must be a lower limit for the mass of black holes. Theoretically, this boundary is expected to lie around the Planck mass ($m_p = \sqrt{\hbar c/G} \approx 1.2 \times 10^{19}$ GeV/$c^2 \approx 2.2 \times 10^{-8}$ kg), where quantum effects are expected to invalidate the predictions of general relativity. This would put the creation of black holes firmly out of reach of any high-energy process occurring on or near the Earth. However, certain developments in quantum gravity suggest that the Planck mass could be much lower: some braneworld scenarios for example put the boundary as low as 1 TeV/c^2. This would make it conceivable for micro black holes to be created in the high-energy collisions that occur when cosmirays hit the Earth's atmosphere, or possibly in the Large Hadron Collider at CERN. These theories are very speculative, and the creation of black holes in these processes is deemed unlikely by many specialists. Even if micro black holes could be formed, it is expected that they would evaporate in about 10^{-25} seconds, posing no threat to the Earth.

Growth

Once a black hole has formed, it can continue to grow by absorbing additional matter. Any black hole will continually absorb gas and interstellar dust from its surroundings and omnipresent cosmibackground radiation. This is the primary process through which supermassive black holes seem to have grown. A similar process has been suggested for the formation of intermediate-mass black holes found in globular clusters.

Another possibility for black hole growth, is for a black hole to merge with other objects such as stars or even other black holes. Although not necessary for growth, this is thought to have been important, especially for the early development of supermassive black holes, which could have formed from the coagulation of many smaller objects. The process has also been proposed as the origin of some intermediate-mass black holes.

Evaporation

In 1974, Hawking predicted that black holes are not entirely black but emit small amounts of thermal radiation; this effect has become known as Hawking radiation. By applying quantum field theory to a statiblack hole background, he determined that a black hole should emit particles that display a perfect black body spectrum. Since Hawking's publication, many others have verified the result through various approaches. If Hawking's theory of black hole radiation is correct, then black holes are expected to shrink and evaporate over time as they lose mass by the emission of photons and other particles. The temperature of this thermal spectrum (Hawking temperature) is proportional to the surface gravity of the black hole, which, for a Schwarzschild black hole, is inversely proportional to the mass. Hence, large black holes emit less radiation than small black holes.

A stellar black hole of $1\ M_\odot$ has a Hawking temperature of about 100 nanokelvins. This is far less than the 2.7 K temperature of the cosmimicrowave background radiation. Stellar-mass or larger black holes receive more mass from the cosmimicrowave background than they emit through Hawking radiation and thus will grow instead of shrink. To have a Hawking temperature larger than 2.7 K (and be able to evaporate), a black hole would need a mass less than the Moon. Such a black hole would have a diameter of less than a tenth of a millimeter.

If a black hole is very small, the radiation effects are expected to become very strong. Even a black hole that is heavy compared to a human would evaporate in an instant. A black hole with the mass of a car would have a diameter of about 10^{-24} m and take a nanosecond to evaporate, during which time it would briefly have a luminosity of more than 200 times that of the Sun. Lower-mass black holes are expected to evaporate even faster; for example, a black hole of mass $1\ \mathrm{TeV}/c^2$ would take less than 10^{-88} seconds to evaporate completely. For such a small black hole, quantum gravitation effects are expected to play an important role and could hypothetically make such a small black hole stable, although current developments in quantum gravity do not indicate so.

The Hawking radiation for an astrophysical black hole is predicted to be very weak and would thus be exceedingly difficult to detect from Earth. A possible exception, however, is the burst of gamma rays emitted in the last stage of the evaporation of primordial black holes. Searches for such flashes have proven unsuccessful and provide stringent limits on the possibility of existence of low mass primordial black holes. NASA's Fermi Gamma-ray Space Telescope launched in 2008 will continue the search for these flashes.

Observational Evidence

Gas cloud ripped apart by black hole at the centre of the Milky Way.

By their very nature, black holes do not directly emit any electromagnetiradiation other than the hypothetical Hawking radiation, so astrophysicists searching for black holes must generally rely on indirect observations. For example, a black hole's existence can sometimes be inferred by observing its gravitational interactions with its surroundings. However, the Event Horizon Telescope (EHT), run by MIT's Haystack Observatory, is an attempt to directly observe the immediate environment of the event horizon of Sagittarius A*, the black hole at the centre of the Milky Way. The first image of the event horizon may appear as early as 2018. The existence of magnetifields just outside the event horizon of Sagittarius A*, which were predicted by theoretical studies of black holes, was confirmed by the EHT in 2015.

Detection of Gravitational Waves from Merging Black Holes

On 24 September 2015 the LIGO gravitational wave observatory made the first-ever successful observation of gravitational waves. The signal was consistent with theoretical predictions for the gravitational waves produced by the merger of two black holes: one with about 36 solar masses, and the other around 29 solar masses. This observation provides the most concrete evidence for the existence of black holes to date. For instance, the gravitational wave signal suggests that the separation of the two object prior to merger was just 350 km (or roughly 4 times the Schwarzschild radius corresponding to the inferred masses). The objects must therefore have been extremely compact, leaving black holes as the most plausible interpretation.

More importantly, the signal observed by LIGO also included the start of the post-merger ringdown, the signal produced as the newly formed compact object settles down to a stationary state. Arguably, the ringdown is the most direct way of observing a black hole. From the LIGO signal it is possible to extract the frequency and damping time of the dominant mode of the ringdown. From these it is possible to infer the mass and angular momentum of the final object, which match independent predictions from numerical simulations of the merger. The frequency and decay time of the dominant mode are determined by the geometry of the photon sphere. Hence, observation of this mode confirms the presence of a photon sphere, however it cannot exclude possible exotialternatives to black holes that are compact enough to have a photon sphere.

The observation also provides the first observational evidence for the existence of stellar-mass black hole binaries. Furthermore, it is the first observational evidence of stellar-mass black holes weighing 25 solar masses or more.

Proper Motions of Stars Orbiting Sagittarius A*

The proper motions of stars near the center of our own Milky Way provide strong observational evidence that these stars are orbiting a supermassive black hole. Since 1995, astronomers have tracked the motions of 90 stars orbiting an invisible object coincident with the radio source Sagittarius A*. By fitting their motions to Keplerian orbits, the astronomers were able to infer, in 1998, that a 2.6 million $M_\odot$ object must be contained in a volume with a radius of 0.02 light-years to cause the motions of those stars. Since then, one of the stars—called S2—has completed a full orbit. From the orbital data, astronomers were able to make refine the calculations of the mass to 4.3 million $M_\odot$ and a radius of less than 0.002 lightyears for the object causing the orbital motion of those stars. The upper limit on the object's size is still too large to test whether it is smaller than its Schwarzschild radius; nevertheless, these observations strongly suggest that the central object is a supermassive black hole as there are no other plausible scenarios for confining so much invisible mass into such a small volume. Additionally, there is some observational evidence that this object might possess an event horizon, a feature unique to black holes.

Accretion of Matter

Black hole with corona, X-ray source (artist's concept).

Due to conservation of angular momentum, gas falling into the gravitational well created by a massive object will typically form a disc-like structure around the object. Artists' impressions such as the accompanying representation of a black hole with corona commonly depict the black hole as if it were a flat-space material body hiding the part of the disjust behind it, but detailed mathematical modelling shows that the image of the diswould actually be distorted by the bending of light that originated behind the black hole in such a way that the upper side of the diswould be entirely visible, while there would be a partially visible secondary image of the underside of the disk.

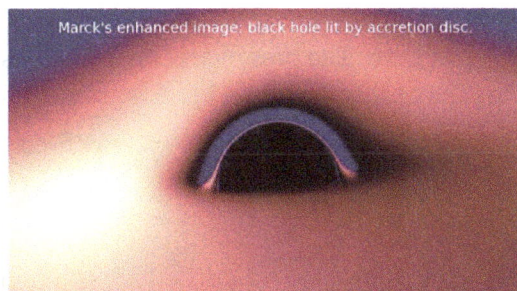

Predicted view from outside the horizon of a Schwarzschild black hole lit by a thin accretion disc

Within such a disc, friction would cause angular momentum to be transported outward, allowing matter to fall further inward, thus releasing potential energy and increasing the temperature of the gas.

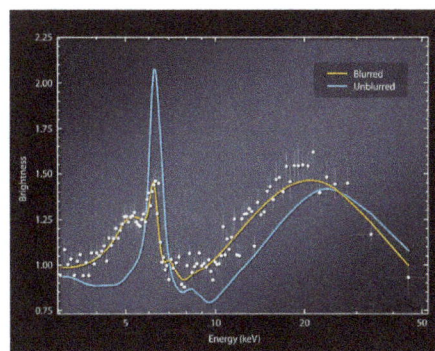

Blurring of X-rays near black hole (NuSTAR; 12 August 2014).

When the accreting object is a neutron star or a black hole, the gas in the inner accretion disorbits at very high speeds because of its proximity to the compact object. The resulting friction is so significant that it heats the inner disto temperatures at which it emits vast amounts of electromagnetiradiation (mainly X-rays). These bright X-ray sources may be detected by telescopes. This process of accretion is one of the most efficient energy-producing processes known; up to 40% of the rest mass of the accreted material can be emitted as radiation. (In nuclear fusion only about 0.7% of the rest mass will be emitted as energy.) In many cases, accretion discs are accompanied by relativistijets that are emitted along the poles, which carry away much of the energy. The mechanism for the creation of these jets is currently not well understood.

As such, many of the universe's more energetiphenomena have been attributed to the accretion of matter on black holes. In particular, active galactinuclei and quasars are believed to be the accretion discs of supermassive black holes. Similarly, X-ray binaries are generally accepted to be binary star systems in which one of the two stars is a compact object accreting matter from its companion. It has also been suggested that some ultraluminous X-ray sources may be the accretion disks of intermediate-mass black holes.

In November 2011 the first direct observation of a quasar accretion disk around a supermassive black hole was reported.

X-ray Binaries

X-ray binaries are binary star systems that emit a majority of their radiation in the X-ray part of the spectrum. These X-ray emissions are generally thought to result when one of the stars (compact object) accretes matter from another (regular) star. The presence of an ordinary star in such a system provides a unique opportunity for studying the central object and to determine if it might be a black hole.

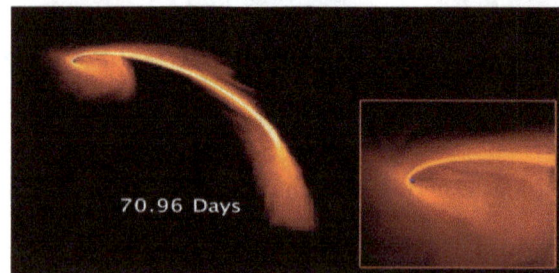

A computer simulation of a star being consumed by a black hole. The blue dot indicates the location of the black hole.

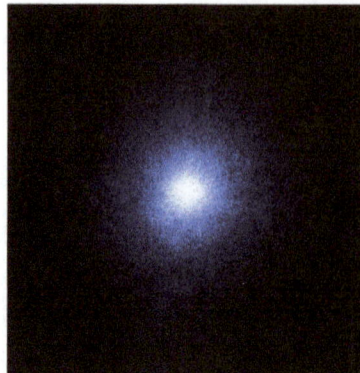

A Chandra X-Ray Observatory image of Cygnus X-1, which was the first strong black hole candidate discovered

This animation compares the X-ray 'heartbeats' of GRS 1915 and IGR J17091, two black holes that ingest gas from companion stars.

If such a system emits signals that can be directly traced back to the compact object, it cannot be a black hole. The absence of such a signal does, however, not exclude the possibility that the compact object is a neutron star. By studying the companion star it is often possible to obtain the orbital parameters of the system and to obtain an estimate for the mass of the compact object. If this is much larger than the Tolman–Oppenheimer–Volkoff limit (that is, the maximum mass a neutron star can have before it collapses) then the object cannot be a neutron star and is generally expected to be a black hole.

The first strong candidate for a black hole, Cygnus X-1, was discovered in this way by Charles Thomas Bolton, Louise Webster and Paul Murdin in 1972. Some doubt, however, remained due to the uncertainties that result from the companion star being much heavier than the candidate black hole. Currently, better candidates for black holes are found in a class of X-ray binaries called soft X-ray transients. In this class of system, the companion star is of relatively low mass allowing for more accurate estimates of the black hole mass. Moreover, these systems are actively emit X-rays for only several months once every 10–50 years. During the period of low X-ray emission (called quiescence), the accretion disis extremely faint allowing detailed observation of the companion star during this period. One of the best such candidates is V404 Cyg.

Quiescence and Advection-Dominated Accretion Flow

The faintness of the accretion disof an X-ray binary during quiescence is suspected to be caused by the flow of mass entering a mode called an advection-dominated accretion flow (ADAF). In this mode, almost all the energy generated by friction in the disis swept along with the flow instead of radiated away. If this model is correct, then it forms strong qualitative evidence for the presence of an event horizon, since if the object at the center of the dishad a solid surface, it would emit large amounts of radiation as the highly energetigas hits the surface, an effect that is observed for neutron stars in a similar state.

Quasi-Periodioscillations

The X-ray emissions from accretion disks sometimes flicker at certain frequencies. These signals are called quasi-periodioscillations and are thought to be caused by material moving along the inner edge of the accretion disk (the innermost stable circular orbit). As such their frequency is linked to the mass of the compact object. They can thus be used as an alternative way to determine the mass of candidate black holes.

Galactinuclei

Astronomers use the term "active galaxy" to describe galaxies with unusual characteristics, such as unusual spectral line emission and very strong radio emission. Theoretical and observational studies have shown that the activity in these active galactinuclei (AGN) may be explained by the presence of supermassive black holes, which can be millions of times more massive than stellar ones. The models of these AGN consist of a central black hole that may be millions or billions of times more massive than the Sun; a disk of gas and dust called an accretion disk; and two jets perpendicular to the accretion disk.

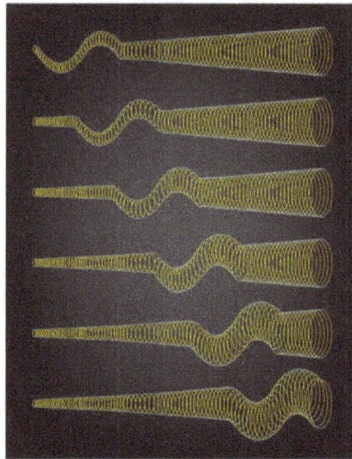

Magnetiwaves, called Alfvén S-waves, flow from the base of black hole jets.

Detection of unusually bright X-Ray flare from Sagittarius A*, a black hole in the center of the Milky Way galaxy on 5 January 2015.

Although supermassive black holes are expected to be found in most AGN, only some galaxies' nuclei have been more carefully studied in attempts to both identify and measure the actual masses of the central supermassive black hole candidates. Some of the most notable galaxies with supermassive black hole candidates include the Andromeda Galaxy, M32, M87, NG3115, NG3377, NG4258, NG4889, NG1277, OJ 287, APM 08279+5255 and the Sombrero Galaxy.

It is now widely accepted that the center of nearly every galaxy, not just active ones, contains a supermassive black hole. The close observational correlation between the mass of this hole and the velocity dispersion of the host galaxy's bulge, known as the M-sigma relation, strongly suggests a connection between the formation of the black hole and the galaxy itself.

Simulation of gas cloud after close approach to the black hole at the centre of the Milky Way.

Microlensing (Proposed)

Another way that the black hole nature of an object may be tested in the future is through observation of effects caused by a strong gravitational field in their vicinity. One such effect is gravitational lensing: The deformation of spacetime around a massive object causes light rays to be deflected much as light passing through an optilens. Observations have been made of weak gravitational lensing, in which light rays are deflected by only a few arcseconds. However, it has never been directly observed for a black hole. One possibility for observing gravitational lensing by a black hole would be to observe stars in orbit around the black hole. There are several candidates for such an observation in orbit around Sagittarius A*.

Alternatives

The evidence for stellar black holes strongly relies on the existence of an upper limit for the mass of a neutron star. The size of this limit heavily depends on the assumptions made about the properties of dense matter. New exotiphases of matter could push up this bound. A phase of free quarks at high density might allow the existence of dense quark stars, and some supersymmetrimodels predict the existence of Q stars. Some extensions of the standard model posit the existence of preons as fundamental building blocks of quarks and leptons, which could hypothetically form preon stars. These hypothetical models could potentially explain a number of observations of stellar black hole candidates. However, it can be shown from arguments in general relativity that any such object will have a maximum mass.

Since the average density of a black hole inside its Schwarzschild radius is inversely proportional to the square of its mass, supermassive black holes are much less dense than stellar black holes (the average density of a $10^8\ M_\odot$ black hole is comparable to that of water). Consequently, the physics of matter forming a supermassive black hole is much better understood and the possible alternative explanations for supermassive black hole observations are much more mundane. For example, a supermassive black hole could be modelled by a large cluster of very dark objects. However, such alternatives are typically not stable enough to explain the supermassive black hole candidates.

The evidence for the existence of stellar and supermassive black holes implies that in order for black holes to not form, general relativity must fail as a theory of gravity, perhaps due to the onset of quantum mechanical corrections. A much anticipated feature of a theory of quantum gravity is that it will not feature singularities or event horizons and thus black holes would not be real artifacts. In 2002, much attention has been drawn by the fuzzball model in string theory. Based on calculations for specifisituations in string theory, the proposal suggests that generically the individual states of a black hole solution do not have an event horizon or singularity, but that for a classical/semi-classical observer the statistical average of such states appears just as an ordinary black hole as deduced from general relativity.

Open Questions

Entropy and Thermodynamics

$$S = \frac{1}{4}\frac{c^3 k}{G\hbar} A$$

The formula for the Bekenstein–Hawking entropy (S) of a black hole, which depends on the area of the black hole (A). The constants are the speed of light (c), the Boltzmann constant (k), Newton's constant (G), and the reduced Planck constant ($\hbar$).

In 1971, Hawking showed under general conditions that the total area of the event horizons of any collection of classical black holes can never decrease, even if they collide and merge. This result, now known as the second law of black hole mechanics, is remarkably similar to the second law of thermodynamics, which states that the total entropy of a system can never decrease. As with classical objects at absolute zero temperature, it was assumed that black holes had zero entropy. If this were the case, the second law of thermodynamics would be violated by entropy-laden matter entering a black hole, resulting in a decrease of the total entropy of the universe. Therefore, Bekenstein proposed that a black hole should have an entropy, and that it should be proportional to its horizon area.

The link with the laws of thermodynamics was further strengthened by Hawking's discovery that quantum field theory predicts that a black hole radiates blackbody radiation at a constant temperature. This seemingly causes a violation of the second law of black hole mechanics, since the radiation will carry away energy from the black hole causing it to shrink. The radiation, however also carries away entropy, and it can be proven under general assumptions that the sum of the entropy of the matter surrounding a black hole and one quarter of the area of the horizon as measured in Planck units is in fact always increasing. This allows the formulation of the first law of black hole mechanics as an analogue of the first law of thermodynamics, with the mass acting as energy, the surface gravity as temperature and the area as entropy.

One puzzling feature is that the entropy of a black hole scales with its area rather than with its volume, since entropy is normally an extensive quantity that scales linearly with the volume of the system. This odd property led Gerard 't Hooft and Leonard Susskind to propose the holographi-principle, which suggests that anything that happens in a volume of spacetime can be described by data on the boundary of that volume.

Although general relativity can be used to perform a semi-classical calculation of black hole entropy, this situation is theoretically unsatisfying. In statistical mechanics, entropy is understood as counting the number of microscopiconfigurations of a system that have the same macroscopiqualities (such as mass, charge, pressure, etc.). Without a satisfactory theory of quantum gravity, one cannot perform such a computation for black holes. Some progress has been made in various approaches to quantum gravity. In 1995, Andrew Strominger and Cumrun Vafa showed that counting the microstates of a specifisupersymmetriblack hole in string theory reproduced the Bekenstein–Hawking entropy. Since then, similar results have been reported for different black holes both in string theory and in other approaches to quantum gravity like loop quantum gravity.

Information Loss Paradox

Because a black hole has only a few internal parameters, most of the information about the matter that went into forming the black hole is lost. Regardless of the type of matter which goes into a black hole, it appears that only information concerning the total mass, charge, and angular momentum are conserved. As long as black holes were thought to persist forever this information loss is not that problematic, as the information can be thought of as existing inside the black hole, inaccessible from the outside. However, black holes slowly evaporate by emitting Hawking radiation. This radiation does not appear to carry any additional information about the matter that formed the black hole, meaning that this information appears to be gone forever.

The question whether information is truly lost in black holes (the black hole information para-dox) has divided the theoretical physics community. In quantum mechanics, loss of information corresponds to the violation of vital property called unitarity, which has to do with the conservation of probability. It has been argued that loss of unitarity would also imply violation of conservation of energy. Over recent years evidence has been building that indeed information and unitarity are preserved in a full quantum gravitational treatment of the problem.

Planets

Some astromomers have suggested that there might be black holes orbited by planets.

Black Hole Information Paradox

Artist's representation of a black hole

The black hole information paradox is a puzzle resulting from the combination of quantum mechanics and general relativity. Calculations suggest that physical information could permanently disappear in a black hole, allowing many physical states to devolve into the same state. This is controversial because it violates a commonly assumed tenet of science—that *in principle* complete information about a physical system at one point in time should determine its state at any other time. A fundamental postulate of quantum mechanics is that complete information about a system is encoded in its wave function up to when the wave function collapses. The evolution of the wave function is determined by a unitary operator, and unitarity implies that information is conserved in the quantum sense.

Principles in Action

There are two main principles in play:

- Quantum determinism means that given a present wave function, its future changes are uniquely determined by the evolution operator.

- Reversibility refers to the fact that the evolution operator has an inverse, meaning that the past wave functions are similarly unique.

The combination of the two means that information must always be preserved.

Starting in the mid-1970s, Stephen Hawking and Jacob Bekenstein put forward theoretical arguments based on general relativity and quantum field theory that not only appeared to be inconsistent with information conservation but were not accounting for the information loss and state no reason for it. Specifically, Hawking's calculations indicated that black hole evaporation via Hawking radiation does not preserve information. Today, many physicists believe that the holographiprinciple (specifically the AdS/CFT duality) demonstrates that Hawking's conclusion was incorrect, and that information is in fact preserved. In 2004 Hawking himself conceded a bet he had made, agreeing that black hole evaporation does in fact preserve information.

Hawking Radiation

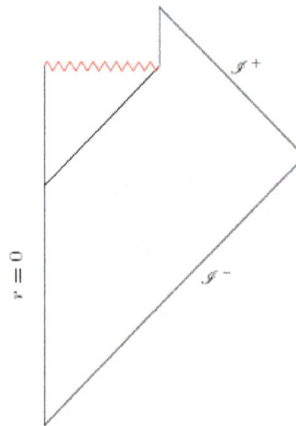

The Penrose diagram of a black hole which forms, and then completely evaporates away. Information falling into it will hit the singularity. Time shown on vertical axis from bottom to top; space shown on horizontal axis from left (radius zero) to right (growing radius).

In 1975, Stephen Hawking and Jacob Bekenstein showed that black holes should slowly radiate away energy, which poses a problem. From the no-hair theorem, one would expect the Hawking radiation to be completely independent of the material entering the black hole. Nevertheless, if the material entering the black hole were a pure quantum state, the transformation of that state into the mixed state of Hawking radiation would destroy information about the original quantum state. This violates Liouville's theorem and presents a physical paradox.

More precisely, if there is an entangled pure state, and one part of the entangled system is thrown into the black hole while keeping the other part outside, the result is a mixed state after the partial trace is taken into the interior of the black hole. But since everything within the interior of the black hole will hit the singularity within a finite time, the part which is traced over partially might disappear completely from the physical system.

Hawking remained convinced that the equations of black-hole thermodynamics together with the *no-hair theorem* led to the conclusion that quantum information may be destroyed. This annoyed many physicists, notably John Preskill, who bet Hawking and Kip Thorne in 1997 that information was not lost in black holes. The implications that Hawking had opened led to a "battle" where Leonard Susskind and Gerard 't Hooft publicly 'declared war' on Hawking's solution, with Susskind publishing a popular book, *The Black Hole War*, about the debate in 2008. (The book carefully notes that the 'war' was purely a scientifione, and that at a personal level, the participants remained friends.) The solution to the problem that concluded the battle is the holographiprinciple, which was first proposed by 't Hooft but was given a precise string theory interpretation by Susskind. With this, "Susskind quashes Hawking in [the] quarrel over quantum quandary".

There are various ideas about how the paradox is solved. Since the 1997 proposal of the AdS/CFT correspondence, the predominant belief among physicists is that information is preserved and that Hawking radiation is not precisely thermal but receives quantum corrections. Other possibilities include the information being contained in a Planckian remnant left over at the end of Hawking radiation or a modification of the laws of quantum mechanics to allow for non-unitary time evolution.

In July 2004, Stephen Hawking published a paper presenting a theory that quantum perturbations of the event horizon could allow information to escape from a black hole, which would resolve the information paradox. His argument assumes the unitarity of the AdS/CFT correspondence which implies that an AdS black hole that is dual to a thermal conformal field theory. When announcing his result, Hawking also conceded the 1997 bet, paying Preskill with a baseball encyclopedia "from which information can be retrieved at will."

According to Roger Penrose, loss of unitarity in quantum systems is not a problem: quantum measurements are by themselves already non-unitary. Penrose claims that quantum systems will in fact no longer evolve unitarily as soon as gravitation comes into play, precisely as in black holes. The Conformal CycliCosmology advocated by Penrose critically depends on the condition that information is in fact lost in black holes. This new cosmological model might in future be tested experimentally by detailed analysis of the cosmimicrowave background radiation (CMB): if true the CMB should exhibit circular patterns with slightly lower or slightly higher temperatures. In November 2010, Penrose and V. G. Gurzadyan announced they had found evidence of such circular patterns, in data from the Wilkinson Microwave Anisotropy Probe (WMAP) corroborated by data from the BOOMERanG experiment. The significance of the findings was subsequently debated by others.

Postulated Solutions

- Information is irretrievably lost

Advantage: Seems to be a direct consequence of relatively non-controversial calculation based on semiclassical gravity.

Disadvantage: Violates unitarity. (Banks, Susskind and Peskin argued that it also violates energy-momentum conservation or locality, but the argument does not seem to be correct for systems with a large number of degrees of freedom.)

- Information gradually leaks out during the black-hole evaporation

Advantage: Intuitively appealing because it qualitatively resembles information recovery in a classical process of burning.

Disadvantage: Requires a large deviation from classical and semiclassical gravity (which do not allow information to leak out from the black hole) even for macroscopiblack holes for which classical and semiclassical approximations are expected to be good approximations.

- Information suddenly escapes out during the final stage of black-hole evaporation

Advantage: A significant deviation from classical and semiclassical gravity is needed only in the regime in which the effects of quantum gravity are expected to dominate.

Disadvantage: Just before the sudden escape of information, a very small black hole must be able to store an arbitrary amount of information, which violates the Bekenstein bound.

- Information is stored in a Planck-sized remnant

Advantage: No mechanism for information escape is needed.

Disadvantage: To contain the information from any evaporated black hole, the remnants would need to have an infinite number of internal states. It has been argued that it would be possible to produce an infinite amount of pairs of these remnants since they are small and indistinguishable from the perspective of the low-energy effective theory.

- Information is stored in a large remnant

Advantage: The size of remnant increases with the size of the initial black hole, so there is no need for an infinite number of internal states.

Disadvantage: Hawking radiation must stop before the black hole reaches the Planck size, which requires a violation of semi-classical gravity at a macroscopiscale.

- Information is stored in a baby universe that separates from our own universe.

Advantage: This scenario is predicted by the Einstein–Cartan theory of gravity which extends general relativity to matter with intrinsiangular momentum (spin). No violation of known general principles of physics is needed.

Disadvantage: It is difficult to test the Einstein–Cartan theory because its predictions are significantly different from general-relativistiones only at extremely high densities.

- Information is encoded in the correlations between future and past

Advantage: Semiclassical gravity is sufficient, i.e., the solution does not depend on details of (still not well understood) quantum gravity.

Disadvantage: Contradicts the intuitive view of nature as an entity that evolves with time.

Recent Developments

In 2014, Chris Adami, a physicist at the Michigan State University claimed to have solved the paradox. Since absolute loss of information is not allowed by quantum physics, Adami argues that the "lost" information is contained in stimulated emission that accompanies the Hawking radiation emitted by black holes. His solution has not been confirmed.

In 2015, Modak, Ortíz, Peña and Sudarsky, have argued that the paradox can be dissolved by invoking foundational issues of quantum theory often referred as the Measurement problem of quantum mechanics. This work was built on an earlier proposal by Okon and Sudarsky on the benefits of Objective collapse theory in a much broader context. The original motivation of these studies was the long lasting proposal of Roger Penrose where collapse of the wave-function is said to be inevitable in presence of black holes (and even under the influence of gravitational field). Experimental verification of collapse theories is an ongoing effort.

Hawking et al. on 5 Jan 2016 proposed new postulates on information moving in and out of a black hole. The 2016 work posits that the information is saved in "soft particles", low-energy versions of photons and other particles that exist in zero-energy empty space.

Black Hole Complementarity

Black hole complementarity is a conjectured solution to the black hole information paradox, proposed by Leonard Susskind and Larus Thorlacius, and Gerard 't Hooft.

Ever since Stephen Hawking suggested information is lost in an evaporating black hole once it passes through the event horizon and is inevitably destroyed at the singularity, and that this can turn pure quantum states into mixed states, some physicists have wondered if a complete theory of quantum gravity might be able to conserve information with a unitary time evolution. But how can this be possible if information cannot escape the event horizon without traveling faster than light? This seems to rule out Hawking radiation as the carrier of the missing information. It also appears as if information cannot be "reflected" at the event horizon as there is nothing special about it locally.

Leonard Susskind proposed a radical resolution to this problem by claiming that the information is both reflected at the event horizon *and* passes through the event horizon and cannot escape, with the catch being no observer can confirm both stories simultaneously. According to an external observer, the infinite time dilation at the horizon itself makes it appear as if it takes an infinite amount of time to reach the horizon. He also postulated a stretched horizon, which is a membrane hovering about a Planck length outside the event horizon and which is both physical and hot. According to the external observer, infalling information heats up the stretched horizon, which then reradiates it as Hawking radiation, with the entire evolution being unitary. However, according to an infalling observer, nothing special happens at the event horizon itself, and both the observer and the information will hit the singularity. This isn't to say there are two copies of the information lying about — one at or just outside the horizon, and the other inside the black hole — as that would violate the no cloning theorem. Instead, an observer can only detect the information at the horizon itself, or inside, but never both simultaneously. Complementarity is a feature of the quantum mechanics of noncommuting observables, and Susskind proposed that both stories are complementary in the quantum sense.

An infalling observer will see the point of entry of the information as being localized on the event horizon, while an external observer will notice the information being spread out uniformly over the entire stretched horizon before being re-radiated. To an infalling observer, information and entropy pass through the horizon with nothing strange happening. To an external observer, the information and entropy is absorbed into the stretched horizon which acts like a dissipative fluid with entropy, viscosity and electrical conductivity. See the membrane paradigm for more details. The stretched horizon is conducting with surface charges which rapidly spread out over the horizon.

Global symmetries don't exist in quantum gravity. Baryon number is violated, but only at very small scales, and the proton has a very long lifetime. But with a short enough time resolution, the proton oscillates between different baryon numbers and the time warping near the horizon magnifies that. Alternatively, the hot temperatures of the stretched horizon cause the proton to decay. But an infalling observer never has time to see the proton decay.

Recently, it appears that black hole complementarity combined with the monogamy of entanglement suggests the existence of a "firewall".

Black Hole Starship

A black hole starship is a theoretical idea for enabling interstellar travel by propelling a starship by creating an artificial black hole and using a parabolireflector to reflect its Hawking radiation. In 2009, Alexander Bolonkin and Louis Crane, Shawn Westmoreland offered and published a paper and book [1 -3] investigating the feasibility of this idea. Their conclusion was that it was on the edge of possibility, but that quantum gravity effects that are presently unknown will either make it easier, or make it impossible.

Although beyond current technological capabilities, a black hole starship offers some advantages compared to other possible methods. For example, in nuclear fusion or fission, only a small proportion of the mass is converted into energy, so enormous quantities of material would be needed. Thus, a nuclear starship would greatly deplete Earth of fissile and fusile material. One possibility is antimatter, but the manufacturing of antimatter is hugely energy-inefficient, and antimatter is difficult to contain. The Crane and Westmoreland paper continues:

> On the other hand, the process of generating a BH from collapse is naturally efficient, so it would require millions of times less energy than a comparable amount of antimatter or at least tens of thousands of times given some optimistifuture antimatter generator. As to confinement, a BH confines itself. We would need to avoid colliding with it or losing it, but it won't explode. Matter striking a BH would fall into it and add to its mass. So making a BH is extremely difficult, but it would not be as dangerous or hard to handle as a massive quantity of antimatter. Although the process of generating a BH is extremely massive, it does not require any new Physics. Also, if a BH, once created, absorbs new matter, it will radiate it, thus acting as a new energy source; while antimatter can only act as a storage mechanism for energy which has been collected elsewhere and converted at extremely low efficiency. (None of the other ideas suggested for interstellar flight seems viable either. The proposal for an interstellar ramjet turns out to produce more drag than thrust, while the idea of propelling a ship with a laser beam runs into the problem that the beam spreads too fast.)

Criteria

According to the authors, a black hole to be used in space travel needs to meet five criteria:

1. has a long enough lifespan to be useful,
2. is powerful enough to accelerate itself up to a reasonable fraction of the speed of light in a reasonable amount of time,

3. is small enough that we can access the energy to make it,

4. is large enough that we can focus the energy to make it,

5. has mass comparable to a starship.

Black holes seem to have a sweet spot in terms of size, power and lifespan which is almost ideal. A black hole weighing 606,000 metritons (6.06×10^8 kg), or roughly the mass of the Seawise Giant (the longest sea-going ship ever built) would have a Schwarzschild radius of 0.9 attometers (0.9×10^{-18} m, or 9×10^{-19} m), a power output of 160 petawatts (160×10^{15} W, or 1.6×10^{17} W), and a 3.5-year lifespan. With such a power output, the black hole could accelerate to 10% the speed of light in 20 days, assuming 100% conversion of energy into kinetienergy. Assuming only 10% conversion into kinetienergy would only take 10 times longer to accelerate to 0.1(10% of the speed of light).

Getting the black hole to act as a power source and engine also requires a way to convert the Hawking radiation into energy and thrust. One potential method involves placing the hole at the focal point of a parabolireflector attached to the ship, creating forward thrust. A slightly easier, but less efficient method would involve simply absorbing all the gamma radiation heading towards the fore of the ship to push it onwards, and let the rest shoot out the back. This would, however, generate an enormous amount of heat as radiation is absorbed by the dish.

References

- Gillispie, C. C. (2000). Pierre-Simon Laplace, 1749–1827: a life in exact science. Princeton paperbacks. Princeton University Press. p. 175. ISBN 0-691-05027-9.

- Israel, W. (1989). "Dark stars: the evolution of an idea". In Hawking, S. W.; Israel, W. 300 Years of Gravitation. Cambridge University Press. ISBN 978-0-521-37976-2.

- Kox, A. J. (1992). "General Relativity in the Netherlands: 1915–1920". In Eisenstaedt, J.; Kox, A. J. Studies in the history of general relativity. Birkhäuser. p. 41. ISBN 978-0-8176-3479-7.

- Shapiro, S. L.; Teukolsky, S. A. (1983). Black holes, white dwarfs, and neutron stars: the physics of compact objects. John Wiley and Sons. p. 357. ISBN 0-471-87316-0.

- Obers, N. A. (2009). Papantonopoulos, Eleftherios, ed. "Black Holes in Higher-Dimensional Gravity". Lecture Notes in Physics. Lecture Notes in Physics. 769: 211–258. arXiv:0802.0519. doi:10.1007/978-3-540-88460-6. ISBN 978-3-540-88459-0.

- Kerr, R. P. (2009). "The Kerr and Kerr-Schild metrics". In Wiltshire, D. L.; Visser, M.; Scott, S. M. The Kerr Spacetime. Cambridge University Press. arXiv:0706.1109. ISBN 978-0-521-88512-6.

- Carr, B. J. (2005). "Primordial Black Holes: Do They Exist and Are They Useful?". In Suzuki, H.; Yokoyama, J.; Suto, Y.; Sato, K. Inflating Horizon of Particle Astrophysics and Cosmology. Universal Academy Press. arXiv:astro-ph/0511743. ISBN 4-946443-94-0.

- McClintock, J. E.; Remillard, R. A. (2006). "Black Hole Binaries". In Lewin, W.; van der Klis, M. Compact Stellar X-ray Sources. Cambridge University Press. arXiv:astro-ph/0306213. ISBN 0-521-82659-4.

- Sparke, L. S.; Gallagher, J. S. (2000). Galaxies in the Universe: An Introduction. Cambridge University Press. Ch. 9.1. ISBN 0-521-59740-4.

- 't Hooft, G. (2001). "The Holographic Principle". In Zichichi, A. Basics and highlights in fundamental physics. Subnuclear series. 37. World Scientific. arXiv:hep-th/0003004. ISBN 978-981-02-4536-8.

- Carlip, S. (2009). "Black Hole Thermodynamics and Statistical Mechanics". Lecture Notes in Physics. Lecture

Notes in Physics. 769: 89. arXiv:0807.4520. doi:10.1007/978-3-540-88460-6_3. ISBN 978-3-540-88459-0.

- Penrose, Roger (1989). "Newton, quantum theory and reality". Three Hundred Years of Gravitation. Cambridge University Press. p. 17. ISBN 9780521379762.

- Susskind, Leonard; Lindesay, James (31 December 2004). An introduction to black holes, information and the string theory revolution: The holographic universe. World Scientific Publishing Company. ISBN 978-981-256-083-4.

- Rees, M. J.; Volonteri, M. (2007). "Massive black holes: formation and evolution". In Karas, V.; Matt, G. Black Holes from Stars to Galaxies—Across the Range of Masses. Cambridge University Press. pp. 51–58. arXiv:astro-ph/0701512. ISBN 978-0-521-86347-6.

- Doeleman, Shep (4 April 2016). "The Event Horizon Telescope: Imaging and Time-Resolving a Black Hole". Physics @ Berkeley. Event occurs at 46:50. Retrieved 8 July 2016.

- Abbott, Benjamin P.; et al. (LIGO Scientific Collaboration and Virgo Collaboration) (11 February 2016). "Tests of general relativity with GW150914". LIGO. Retrieved 12 February 2016.

Permissions

We would like to thank the editorial team for lending their expertise to make the book truly unique. They have played a crucial role in the development of this book. Without their invaluable contributions this book wouldn't have been possible. They have made vital efforts to compile up to date information on the varied aspects of this subject to make this book a valuable addition to the collection of many professionals and students.

This book was conceptualized with the vision of imparting up-to-date and integrated information in this field. To ensure the same, a matchless editorial board was set up. Every individual on the board went through rigorous rounds of assessment to prove their worth. After which they invested a large part of their time researching and compiling the most relevant data for our readers.

The editorial board has been involved in producing this book since its inception. They have spent rigorous hours researching and exploring the diverse topics which have resulted in the successful publishing of this book. They have passed on their knowledge of decades through this book. To expedite this challenging task, the publisher supported the team at every step. A small team of assistant editors was also appointed to further simplify the editing procedure and attain best results for the readers.

Apart from the editorial board, the designing team has also invested a significant amount of their time in understanding the subject and creating the most relevant covers. They scrutinized every image to scout for the most suitable representation of the subject and create an appropriate cover for the book.

The publishing team has been an ardent support to the editorial, designing and production team. Their endless efforts to recruit the best for this project, has resulted in the accomplishment of this book. They are a veteran in the field of academics and their pool of knowledge is as vast as their experience in printing. Their expertise and guidance has proved useful at every step. Their uncompromising quality standards have made this book an exceptional effort. Their encouragement from time to time has been an inspiration for everyone.

The publisher and the editorial board hope that this book will prove to be a valuable piece of knowledge for students, practitioners and scholars across the globe.

Index